T0177221

The Struggle of Parts

The Struggle of Parts

WILHELM ROUX

TRANSLATED AND EDITED BY

DAVID HAIG AND RICHARD BONDI

Harvard University Press

CAMBRIDGE, MASSACHUSETTS & LONDON, ENGLAND   2024

First printing

Library of Congress Cataloging-in-Publication Data

Names: Roux, Wilhelm, 1850–1924, author. | Haig, David, 1958– editor and
translator. | Bondi, Richard, 1963– editor and translator.
Title: The struggle of parts / Wilhelm Roux ; translated and edited by
David Haig and Richard Bondi.
Other titles: Kampf der Theile im Organismus. English
Description: Cambridge, Massachusetts ; London, England : Harvard
University Press, 2024. | Includes bibliographical references and index.
Identifiers: LCCN 2023033357 | ISBN 9780674290648 (cloth)
Subjects: LCSH: Evolution (Biology) | Variation (Biology) | Morphogenesis.
Classification: LCC QH366 .R785 2024 | DDC 576.8—dc23/eng/20231101
LC record available at https://lccn.loc.gov/2023033357

Contents

Note to the Reader

Square brackets in the translated text indicate text added by Roux in the second edition. These square brackets appear in the original German second-edition text. Curly brackets in the translated text indicate text that appeared in the first edition but was removed from the second edition. Translator's notes are marked with "—Trans." Roux's original notes are unmarked. Translators / editors have excised most of Roux's original citations, only keeping a few where they have particular historiographical relevance. See the introduction by David Haig and Richard Bondi for an explanation.

The Struggle of Parts

Creative Struggle within the Organism

DAVID HAIG AND RICHARD BONDI

On April 16, 1881, Darwin wrote to his friend and disciple George Romanes, "Dr. Roux has sent me a book just published by him *Der Kampf der Theile.* . . . It is full of reasoning, and this in German is very difficult to me, so that I have only skimmed through each page; here and there reading with a little more care. As far as I can imperfectly judge, it is the most important book on Evolution which has appeared for some time" (1887, 3:244).

DESPITE HIS IMPERFECT GERMAN, Darwin discerned that Roux had brought the "struggle for existence" into the body as an internal struggle among molecules, cells, and tissues. In his letter of April 16, 1881, Darwin offered to send his copy of the book to Romanes with the hope that Romanes would write a review for *Nature.* On August 7, 1881, Darwin prodded Romanes, sending him a foreign review: "I received yesterday the enclosed notice, and I send it to you, as I have thought that if you notice Dr. Roux's book in *Nature* or elsewhere the review might possibly be of use to you. As far as I can judge the book ought to be brought before English naturalists." Romanes responded the next day: "Many thanks for the notice of Roux's book. I have not yet looked at the latter, but Preyer, of Jena (who has been our guest during the Congress meeting, and who knows the author), does not think much of it" (Romanes 1896, 120).

Darwin wrote to Romanes again on September 2, 1881: "Two or three papers by Hermann Müller have just appeared in *Kosmos,* which seem to me interesting, as showing how soon, i.e. after how many attempts, bees learn how best to suck a new flower; there is also a good and laudatory review of Dr. Roux." Romanes responded two days later: "I have already sent in a short review of Roux's book, but should like to see about the bees in *Kosmos*." Romanes had read the book and written a review in less than a month, which was published in *Nature* on September 29, 1881.

Romanes's review, "The Struggle of Parts in the Organism" (1881a), consisted of four paragraphs: the first was an obituary for natural theology and a panegyric to Darwin—"Then with a suddenness only less surprising than its completeness the end came; the fountains of this great deep were broken by the power of one man, and never in the history of thought has a change been effected of a comparable magnitude or importance." Romanes's second paragraph argued that natural selection, which the English philosopher Herbert Spencer had called "indirect equilibration," was inadequate to explain adaptation and that there must be additional adaptive processes within organisms, which Herbert Spencer had called "direct equilibration." Roux's book was an "interesting effort" in the direction of understanding Spencer's direct equilibration. The remaining two paragraphs damned the book with faint praise. "Perhaps the most striking feature in the detailed exposition which the author gives of the doctrine is his ignorance of the fact that the doctrine is not original . . . in this country, at all events, the idea is far from being a novel one" (1881a, 505–506).

On October 13, 1881, *Nature* published Romanes's enthusiastic review (1881b) of Darwin's *Formation of Vegetable Mould* (1881). Romanes appears to have interrupted reviewing Darwin's book to write a hasty review of Roux in response to Darwin's persistent reminders. Darwin died on April 19, 1882, so we do not know how he would have pursued his interest in the struggle of parts. Romanes later found Roux's struggle of parts useful in explaining the degeneration of organs that had fallen into disuse (Romanes 1895).

Der Kampf der Theile im Organismus (1881) had another reader, fluent in German, who was unlikely to have been sent a presentation copy by Roux. Friedrich Nietzsche acquired a copy and made copious notes on it during 1881 (Moore 2002, 78). He interpreted the book as an affirmation of internal conflict and a diminution of the importance of the external Darwinian struggle for existence (Müller-Lauter 1999). In a passage from his notebooks entitled "*Gegen den Darwinismus*" (Against Darwinism) Nietzsche wrote, "The individual itself as a struggle of parts (for nourishment, space, etc.); its development bound to a victory over and complete dominance of individual parts; to a withering, an 'organ-becoming,' of other parts. The influence of 'external circumstances' is absurdly overestimated by Darwin; because the essence of the life-process is precisely this stupendous formative violent power, hewing form from within, using, exploiting, the 'external circumstances'" (Nietzsche 1906, §647; 2017 367).

THREE READERS. Three opinions. Why had Darwin considered *Der Kampf der Theile* an important extension of his own work? Why was Romanes unimpressed? Why did Nietzsche interpret the book as opposed to Darwinism? These have been difficult questions for monoglot English readers to answer because Roux's treatise has, until now, never been translated into English. We hope our translation will help you answer these questions by making his ideas accessible to readers who, like Charles Darwin, struggle with German syntax and grammar.

Wilhelm Roux

Wilhelm Roux was born in Jena in 1850, a subject of the grand-duke of Sachsen–Weimar–Eisenach. He died in Halle in 1924, a citizen of the Weimar Republic. His life thus encompassed the rise of the second German Empire in martial glory and its fall in ignominious defeat. At the time of his birth, Sachsen–Weimar–Eisenach was part of the loose German Confederation, which had replaced the Holy Roman Empire after the defeat of Napoleon. The grand duchy remained neutral in the Austro-Prussian War of 1866 but joined the Prussian-dominated North German Confederation after Austria's defeat and became part of the German Empire after France's defeat in the Franco-Prussian war of 1870–1871.

These wars of German unification—orchestrated by Bismarck and culminating in the proclamation of a new German empire—were the major political events of Roux's youth. When he published *Der Kampf der Theile* in 1881, Germany was a recently unified, rising power. Roux lived through

World War I as a subject of the Kaiser and director of the Anatomical Institute in Halle, where he experienced the Kaiser's forced abdication and Germany's surrender in November 1918. He died after the defeat of an attempt to establish a German socialist republic but before the rise to power of the National Socialists under Adolf Hitler.

Wilhelm Roux as a Student

Fencing was the family business: Wilhelm Roux's father, uncles, grandfather, and great-grandfather had all been fencing masters at universities and institutes across politically fragmented Germany. At the time of his birth, Wilhelm's father, Wilhelm Ludwig Roux, was fencing master at the University of Jena and the author of an influential fencing manual. Wilhelm Ludwig also trained his sons for the family profession. His older son, Ludwig Caesar Roux, was appointed fencing master at the university of Leipzig in 1865, and the father had similar hopes for his younger son. Wilhelm had other plans. He later described his youth as joyless and himself as withdrawn and quiet. He wanted to study medicine but had been sent to a *Realschule* rather than *Gymnasium*. Without a gymnasium qualification, Wilhelm could not enroll in the university's faculty of medicine; as a stop-gap, he enrolled in its faculty of philosophy at Easter of 1870.

Wilhelm was called for military service when the North German Confederation mobilized in July 1870 as a prelude to the outbreak of the Franco-Prussian War later that year. After the German victory and annexation of Elsass (Alsace), Wilhelm Ludwig Roux wanted his son to become fencing master at the new German university of Straßburg (the once and future French university of Strasbourg), and he sent a letter of recommendation to its rector. Fortunately for Wilhelm, the letter was misdirected, and he was able to continue his studies at Jena while studying for the gymnasium examinations, which he passed the following year. He enrolled in the faculty of medicine on January 8, 1873, and passed his *Physikum* in 1874.

Jena was a minor German university. Successful faculty left Jena for prestigious appointments elsewhere, but when Wilhelm was enrolled as a student in 1870 it was the center of German *Darwinismus* in the guise of an "evolutionary morphology" taught by Karl Gegenbaur, professor of anatomy in the faculty of medicine, and Ernst Haeckel, professor of zoology in the faculty of philosophy. Gegenbaur's research focused on the comparative anatomy of vertebrates whereas Haeckel focused on the embryology of invertebrates. Both used comparative morphology to infer lines of evolutionary descent. When Gegenbaur left Jena for Heidelberg in 1873, he was replaced by Gustav Schwalbe as professor of anatomy. The professor of physiology in the faculty of medicine during the 1870s was William Preyer, later a founder of child psychology (and the house guest of George Romanes in 1881).

In the foreword to the second edition of *Der Kampf der Theile*, Roux recalled attending lectures given by Haeckel, Gegenbauer, and Preyer. During 1876 he spent two semesters with the pathologist Rudolf Virchow in Berlin, widely considered one of the world's leading experts in medicine and public health. Roux then returned to Jena before spending his final semester in Straßburg where he attended lectures by Friedrich von Recklinghausen, another pathologist. Roux passed the state medical examination in Jena in the winter of 1877–1878. His doctoral thesis, "On the Ramifications of the Blood Vessels," supervised by Schwalbe, was accepted in April 1878.

Wilhelm Roux as a Scientist

Now that his son was medically qualified, Wilhelm Ludwig Roux hoped Wilhelm would become a general practitioner, but Wilhelm instead took a low-paying job at Franz Hofmann's Institute of Hygiene in Leipzig to learn methods of chemical analysis and experimental procedure. From this institutional base, Roux published "On the Causes of the Deflection of the Arterial Trunk during the Branching Process" (1879) in which he briefly presented his ideas on the struggle of parts and functional adaptation, and mentioned his regret that he was unable to continue this work. This article resulted in an invitation from Carl Hasse, professor of anatomy at the University of Breslau (now Wroclaw in Poland), to become an assistant in Hasse's Anatomical Institute from October 1, 1879.

Roux habilitated in Breslau on August 1, 1880, on the basis of a lecture and an excerpt from what was to become the first chapter of *Der Kampf der Theile*. His habilitation gave Roux the academic status of *Privatdozent* (unsalaried teaching assistant). He then rapidly completed *Der Kampf der*

Theile, which was published in February 1881. In the year before his death, Roux reminisced, "This theory took shape in my brain all by itself, at least from the germs of thought that I had as a student. At all times of the day and night I wrote down the series of thoughts that were arising on the sheets of paper that were always available, and when I set down my short-hand pen I often did not know what we had both written" (1923a, 106).

Roux is better known as a pioneer and polemicist of *Entwicklungs-mechanik* (developmental mechanics) in work published after *Der Kampf der Theile.* When *Der Kampf der Theile* is mentioned in histories of embryology, it is often briefly alluded to as a juvenile work written while Roux was still in thrall to Haeckel and before he had shifted his focus from evolutionary questions to proximate mechanisms of development (e.g., Maienschein 1994; Hamburger 1997). It is appropriate therefore to briefly discuss Haeckel's evolutionary theories before turning to the contents of *Der Kampf der Theile.*

Haeckel attempted to unite cell theory with the new evolutionary thinking. He believed in a hierarchy of individuality with autonomous cellular parts (plastidules) at the bottom of the hierarchy, then cells, followed by an ascending series of bodily parts before arriving at persons and colonies of organisms. Sometimes his flexible concept of individuality extended to species and phyla. Haeckel's "biogenetic law" stated that the development of an individual (ontogeny) recapitulates the evolutionary history of its species (phylogeny). Under this law, the ontogenetic development of an organism reflected the phylogenetic history of its evolutionary lineage. In formulation of this law, Haeckel coined the terms *Ontogenie* and *Phylogenie* to distinguish different aspects of morphological change that had been previously subsumed under the general title of *Entwicklung* (development).

Der Kampf der Theile adopted a version of Haeckel's hierarchy of individuality and posited a struggle for existence among individual parts at each level of the hierarchy. In Roux's view, adaptive outcomes of struggles within an organism during its development could become heritable features of subsequent generations via the inheritance of acquired characters. This process could account for many of the purposive features of organisms. For Haeckel, individual development replayed phylogenetic history and provided clues for reconstructing that history. Phylogenetic changes were incorporated into individual ontogeny. Roux reversed the emphasis: changes during individual development were incorporated into phylogenetic change. Roux saw this as providing a mechanistic explanation of functional adaptation and of organismal purposiveness. He further argued that the physical forces and functional stimuli to which cells and tissues were subject shaped their mature form. Thus, *Der Kampf der Theile* looks backward toward Haeckelian theories of evolutionary change and forward toward the mechanical causes of development that were to be the focus of *Entwicklungsmechanik*.

Roux coined the term *Entwicklungsmechanik* around 1884 to describe his vision of a causal, rather than merely descriptive, science of embryology. In 1885, he kept the medullary plate of a chicken embryo alive in saline solution for thirteen days (sometimes considered the beginning of tissue culture) and reported a series of experiments in which half the cells were killed in early frog embryos (in Roux's own opinion, the beginning of the study of causal embryology). These embryos developed into half a frog, from which Roux concluded that the developmental fate of cells had already been determined.

Roux was appointed associate professor of anatomy in Breslau in 1886. On July 27, 1888, he became director of a new institute of *Entwicklungsmechanik* in Breslau but remained in that position for barely a year. On August 23, 1889, he became professor of anatomy at Innsbruck in Austria, where he remained for the next six years. In 1894, he founded the journal *Archiv für Entwicklungsmechanik* (still published today under the title *Development, Genes and Evolution*). On August 15, 1895, he was appointed director of the Anatomical Institute at Halle in Prussian Saxony and remained in that position until his mandatory retirement in 1921. He died in Halle in 1924.

Our major source for this account of Roux's life is his *Selbstdarstellung* (self-portrayal) published the year before his death (1923a). The self-portrayal focuses on science and says little about Roux's personal life or the wars through which he lived. This account was supplemented by Frederick Churchill's (1994) brief biography of Roux and Lynn Nyhart's (1995) account of the origins of evolutionary morphology at Jena. Alex Kiermayer (2020) was consulted for the history of the Roux family of fencing masters.

Wilhelm Roux as a Philosopher

Wilhelm Roux studied philosophy for a year under Rudolf Eucken (winner of the 1908 Nobel Prize for Literature) while completing his doctoral thesis at Jena (1923a, 145f). He fondly recalled late nights at Eucken's home, debating philosophical questions over a glass of beer. These conversations undoubtedly influenced the writing of *Der Kampf der Theile* which contains explicit references to Empedocles, Heraclitus, Leucippus, Democritus, and Aristotle (notably not Plato) but does not mention any modern philosophers by name.

Roux wrote that he learned Kant's definition of mechanistic events as "strictly lawful" from Eucken (1923a, 146), and it was in this sense that he intended mechanics to be interpreted in *Entwicklungsmechanik*. The subtitle of the first edition of *Der Kampf der Theile* was *Ein Beitrag zur Vervollständigung der mechanischen Zweckmässigkeitslehre*, which can be translated as "A contribution to the completion of a mechanical theory of purposiveness." The subtitle alludes to Kant's antinomy of teleological judgment. Kant presented this antinomy as a conflict between the first and second maxims of the power of judgment: the *thesis*, "All generation of material things and their forms must be judged as possible in accordance with merely mechanical laws," and the *antithesis*, "Some products of material nature cannot be judged as possible according to merely mechanical laws (judging them requires an entirely different law of causality, namely that of final causes)" ([1790] 2000, §70, 258–259). Because of the limited nature of our cognitive faculties, Kant believed, we would never be able to explain the purposiveness of living things in purely mechanical terms.

In a famous passage, Kant wrote, "For it is quite certain that we can never adequately come to know the organized beings and their internal possibility in accordance with merely mechanical principles of nature, let alone explain them; and indeed this is so certain that we can boldly say that it would be absurd for humans even to make such an attempt or to hope that there may yet arise a Newton who could make comprehensible even the generation of a blade of grass according to natural laws that no intention has ordered; rather, we must absolutely deny this insight to human beings" ([1790] 2000, §75, 270–271).

Kant resolved the antinomy of teleological judgment by arguing that we should accept both thesis and antithesis ([1790] 2000, §78). We have no choice but to see living things as both purposive and mechanical without reduction of the former to the latter. Roux believed he had achieved what Kant said could not be done by showing how purposiveness could arise mechanically without teleology. Perhaps Roux saw himself as deserving the title "Newton of the Grass Blade." *Zweckmäßigkeit* (purposiveness) was really *Dauerfähigkeit* (lastingness). What appeared purposeful was what was left behind after the less-lasting had been eliminated. Roux ascribed this idea to Empedocles. He saw Charles Darwin's and Alfred Russel Wallace's hypothesis of the struggle for existence as a modern exemplar of this ancient idea.

Roux's philosophical interests ran counter to the reigning ideology among natural scientists that philosophy did not belong in rigorous science. This ideology was a reaction to what were seen as the excesses of *Naturphilosophie* in early nineteenth-century biology and the pretensions to scientific authority of Hegelian philosophy within university faculties. In this struggle for authority, Hermann Helmholtz (1865, 7) wrote that philosophers accused natural scientists of narrow-mindedness (*Bornierteit*), scientists accused philosophers of meaninglessness (*Sinnlossigkeit*), and many scientists excluded all philosophical influences from their work. Karl von Baer (1866), a distinguished embryologist from an older generation, decried the pervasive "teleophobia" among younger scientists. Helmholtz, together with Carl Ludwig, Emil du Bois-Reymond, and Ernst Brücke, constituted the so-called Group of 1847, who had vowed to exclude anything but physicochemical causes from physiology (Cranefield 1957; Lenoir 1981). Their program reduced scientific explanation to Aristotle's material and efficient causes with the rejection of his formal and final causes.

After publication of *Der Kampf der Theile,* Roux's former doctoral advisor, Gustav Schwalbe, told Roux he should never again publish such a philosophical work if he wanted to become a professor of anatomy (Roux 1923a, 152). Perhaps Roux's preoccupation with explaining *Zweckmäßigkeit* had made him guilty of the sin of teleology. These criticisms may explain why Roux modified the subtitle of the second edition to *Ein Beitrag zur Vervollständigung der Lehre von der mechanischen Ent-stehung des sogenannten "Zweckmässigen"* (A contribution to the comple-tion of a theory of the mechanical genesis of the so-called "purposive"). In his later career, Roux was a strong critic of the re-emergence of teleology, which he thought he had explained mechanically, in various vitalistic theories of development.

Roux has a reputation as a strict empiricist and fierce advocate of the experimental method, but most of his publications address foundational rather than empirical questions. One of his last published papers appeared in the *Annalen der Philosophie* (1923b). Roux clearly saw his work as con-tributing to philosophy, but we have found no evidence that he knew of his most important influence in the history of philosophy. Friedrich Nietzsche, as we have seen, read *Der Kampf der Theile* soon after its publication in 1881. Gregory Moore (2002) and James Pearson (2023) have argued that Nietzsche's central concept of will to power, in which a struggle for domi-nance among opposing drives forged a higher unity within the self, was strongly influenced by his reading of *Der Kampf der Theile.*

One senses a tension within Roux between the experimentalist and the philosopher. His valedictory self-portrayal begins with a couplet (Roux 1923a):

> *Das Was erforsche, mehr erforsche Wie*
> *Und das Warum versäume nie*

> [Research the What, even more research the How
> And never neglect the Why]

The first line is adapted from a speech by Homunculus in Goethe's *Faust Part II,* with *erforsche* (research) substituted for *bedenke* (consider). The second line is Roux's addition. Roux's self-portrayal ends: "In my en-deavors I believe that I have always placed a question mark at the limit of the current state of exact research and that I have left complete freedom to the 'hypothesis' for the area of the still unknown, but only to the 'hypoth-esis,' not the 'assertion' of the unproven and of the unprovable (mystical)" (1923a, 202).

About the Translation

The first problem we faced was deciding which version to translate. Roux's treatise *Der Kampf der Theile im Organismus* was published at Leipzig in 1881. A shorter essay, also titled "*Der Kampf der Theile im Organismus,*" was published the same year in *Biologisches Centralblatt.* Roux republished the treatise and essay in his *Collected Works* of 1895, as second editions with modified titles. Our translation is based on the second edition of the treatise and thus complements the recent French translation (Roux 2016), which is based on the first edition of the treatise. Roux expressed the hope that he might still produce a new edition of *Der Kampf der Theile* (1923a, 168), and perhaps revisions for a third edition exist in some archive.

The second edition contained a new foreword in which Roux discussed the reception of the first edition. In this foreword, Roux says that the first edition was written rapidly and that he had often given the beginning and end of an argument without the middle. These defects were, to some extent, addressed in his revisions for the second edition but not completely so. Roux can be difficult to read, even for a native German speaker, because he often belabors a point, in different ways and emphasizing different aspects, as if to ensure the reader will not misunderstand; and his sentences, perhaps to prove his intellectual bona fides, can be long and cumbersome. Our translation attempts to clarify Roux's meaning without smoothing all rough edges to preserve some of his awkward style.

Roux's second edition contains extensive footnotes that sometimes take up most of the page and run on to the next page. We have not included these footnotes except where we found a footnote helped to explicate the main text or to place the text in historical context. We have also added some clarifying footnotes of our own. The deletion of most footnotes means that we have not included Roux's citations. Roux's practices of citation were somewhat casual. Sometimes he mentions someone's work without giving a source or mentions names from a secondary source. Scholars who are interested in the citations and footnotes should consult the original German. When Roux mentions another scientist by surname, we have added forenames to make individuals more easily identifiable. For example, at one point Roux refers to "*Newton und nach ihm Obolensky,*" and we have written "Auguste Nélaton (and after him Ivan Nikolaevich Obolensky)." Roux's source is a paper by "J. Obolensky," which discusses work by Nélaton. We have corrected the mistranscription of Nélaton (present in both editions of *Der Kampf der Theile*) and identified which of several possible Obolenskys probably wrote the paper (interpreting the initial J as Johann, the German equivalent of Ivan). We correct a few similar miscitations by Roux.

In the second edition, Roux reproduced most of what he wrote in the first edition even when he had subsequently changed his mind. Substantially new text is placed in [square brackets], sometimes negating the immediately preceding sentence. We have retained the square brackets to show these revisions. Roux sometimes inserts "[?]," which we interpret as one of his ways of expressing reservations about something he had written in the first edition. Roux also reworded parts of the text in the second edition without placing his revisions in square brackets because he considered these changes simple clarifications rather than new material. Occasionally, we detected text from the first edition that had been deleted in the second edition. We have restored the earlier text by placing it in {curly brackets}, but we did not undertake a systematic search for such deletions.

The two editions use italic, bold, or widely spaced fonts as forms of emphasis with considerable differences between the two editions. In some cases, where Roux thought he had been misunderstood, he reprints what had appeared as unemphasized text in the first edition in widely spaced font in the second edition. Our translation does not re-create Roux's use of emphasis because we believe it distracts more than it clarifies. We make sparing use of *italics* to clarify the stream of his argument.

Translation is an art of compromise, not an exact science. The remainder of this section will discuss some of our translation choices to give an idea of our philosophy of translation, to indicate historical changes in the meanings of words, and to flag some words we found particularly troublesome. The titles and subtitles are a good place to begin. *Kampf* can be translated as fight, struggle, combat, or battle. George Romanes (1881a) translated *Der Kampf der Theile* as "The Struggle of Parts" and this has been the standard English translation ever since. We also translated *Kampf* as "struggle" rather than "battle" for historical continuity and because Darwin's phrase "the struggle for existence" appears in Roux and elsewhere as *der Kampf ums Dasein*. We note that German *Kampf* has more martial overtones than English *struggle* and that Roux frequently refers to *Sieg* (victory) and *Herrschaft* (dominance, dominion, rule, hegemony) in the struggle of parts. These militaristic connotations are present, but not strongly emphasized, in our translation.

On the Origin of Species (Darwin 1859) was first translated into German in 1860 by Heinrich Georg Bronn as *Über die Entstehung der Arten*. Bronn used *natürliche Züchtung* for "natural selection" but *sexuelle Zuchtwahl* for "sexual selection." *Züchtung* was a general term for cultivation or breeding whereas *Zuchtwahl* (breeding choice) was a neologism. Gliboff (2008) has argued that Bronn chose *natürliche Züchtung* to emphasize the analogy to agricultural practices but to downplay the connotation that nature "chooses." Bronn nevertheless tolerated the implication of choice in *sexuelle Zuchtwahl* because one individual chooses to mate with another. After Bronn's death, Julius Victor Carus assumed responsibility for subsequent German translations of the *Origin of Species* and used *Zuchtwahl* for both natural and sexual selection. Roux used Carus's translation when writing *Der Kampf der Theile*.

Roux uses a variety of nouns for selective processes. *Zuchtwahl* is only used when directly referring to Darwin's theory (including *natürliche Zuchtwahl* and *geschechtliche* [sexual] *Zuchtwahl*) or quoting from others who have used *Zuchtwahl*. The English loan word *Selection* appears only in compound nouns (*Selectionsprincip, Selectionstheorie, Selectionslehre*). The two words Roux commonly uses are *Züchtung* (but never *natürliche Züchtung* or *geschechtliche Züchtung*) and *Auslese* (including *geschechtliche Auslese*). We chose to translate *Zuchtwahl* and *Auslese* as "selection" but *Züchtung* as "cultivation." We justify the latter choice below.

At the time of writing *Der Kampf der Theile*, *Züchtung* could refer to propagation and cultivation of plants, breeding of animals, or culturing of bacteria. *Selbstzüchtung* could be a program of human self-improvement and, later, a program of eugenical "self-breeding." (*Selbstzucht* was self-discipline.) At the time of the first edition, Roux believed in the inheritance of acquired characters and probably did not see a clear distinction between improvements achieved by nurture or breeding. For this reason, and to distinguish his uses of *Züchtung* and *Auslese,* we decided to translate *Züchtung* as "cultivation." Roux occasionally refers to *innere Umzüchtung.* In the 1880s, *Umzüchtung* was used for the repeated culturing of bacteria, often with the aim of attenuating their virulence. We translate *innere Umzüchtung* as "internal recultivation." The two uses of *Selbstzüchtung* are translated as "self-cultivation." We translate the adjectives *auslesende* as "selective" (the verb *auslesen* is to select or pick over) and *züchtende* as "cultivating."

The subtitle of the first edition states that it is a contribution toward a mechanical *Zweckmässigkeitslehre* (*Lehre* is a teaching, theory, or doctrine). We translate *Zweck* as "purpose," *zweckmässig* as "purposive," and *Zweckmässigkeit* as "purposiveness."

The eponymous hero of Goethe's *Faust* declares, "*Was du ererbt von deinen Vätern hast, erwirb es, um es zu besitzen*" (I:682/3). This can be roughly translated as "What you have inherited from your fathers, earn it, to make it yours." Our reason for quoting Goethe's aphorism is that two of its verbs, *ererben* and *erwerben,* (almost) appear in the phrase "*der Vererbung erworbener Eigenschaften.*" This is a German version of "the inheritance of acquired characters" (English) or "*l'hérédité des caractères acquis*" (French). Versions of this phrase can be found in French from at least the 1830s, but there is a major increase in the frequency of its use in all three languages in the early 1880s. In his versions of the phrase, Roux uses *Vererbung, Vererblichkeit,* and *Vererbbarkeit.* Thus, he consistently uses nouns related to *vererben* (bequeath) rather than *ererben* (inherit). The standard English translation of *Vererbung* has been "inheritance," but we translate it as "bequeathal" because Roux uses *Erblichkeit* for "heritability" and sometimes uses both *ererben* and *vererben* in passages in which he clearly recognizes a distinction between giving (bequeathal) and receiving (inheritance). A second thing to note is that *erworbener* is derived from the past tense of *erwerben,* which Goethe uses in the previous quotation in the sense of "earn." We use the standard translation of *erworbener* as "acquired" but note that *erwerben* has stronger connotations of acquisition by one's own efforts than does the English word.

The standard English translation of *Stoffwechsel* is "metabolism," a word that began to be used by English physiologists in the 1870s, decades after German physiologists had introduced *Stoffwechsel.* We translate

Stoffwechsel as "material exchange" because Roux uses the term with a wider application than the current concept of metabolism. Roux, for example, discusses *Stoffwechsel* of a flame. We translate *Selbstgestaltung* as "self-organization" to indicate continuity with current interest in self-organization, although "self-shaping" might be closer to German etymology (*Gestalt* is notoriously difficult to translate).

Roux posits a hierarchy of struggles among living entities. At the lowest level is a struggle among molecules (*Kampf der Molekeln*). We translate *Molekel* as molecule although Roux's intended meaning is different from the modern chemical concept of a molecule. Roux's *Molekeln* are the ultimate organic process-units of which organisms are composed. They are living things capable of assimilation and multiplication. In the second edition, Roux usually changes the unqualified *Molekel* of the first edition to *lebensthätiges Molekel*, which we translate as "living molecule."

Roux refers to evolutionary theory as *Descendenzlehre,* which we translate as doctrine of descent. We translate the noun *Entwickelung* as "development." German morphologists of this period, including Roux, used *Entwickelung* (or the more modern *Entwicklung*) both for changes in the form of individuals over the course of development and for changes in the form of species over paleontological time. When Roux uses the word *Evolution,* he is always referring to an unfolding during ontogenetic development of something that was preformed as opposed to *Epigenesis,* the formation of form from the formless. We translate *Evolution* as "evolution" but indicate here that the word should be understood in this older developmental sense as synonymous with "preformation."

Roux uses two abstract nouns, *Dauerfähigkeit* and *Leistungsfähigkeit,* to refer to complementary aspects of organic aptness: *Dauerfähigkeit* emphasizes dogged persistence whereas *Leistungsfähigkeit* emphasizes effective action in the world. *Dauer* is duration; *dauerfähig* is lasting, enduring, or permanent. *Dauerfähigkeit* could be translated as durability, permanence, or robustness, but we chose "lastingness" because Roux's emphasis is on a thing's remaining behind after everything else has perished. The verb *leisten* means to achieve, perform, or accomplish (we chose to translate it as "achieve"). The noun *Leistung* could be translated in many ways depending on context including achievement, performance, accomplishment, or output. We usually translate *Leistung* as "achievement" to emphasize the etymological link with *leisten*. *Leistungsfähigkeit* (literally "achievingness," the extent of the capacity to achieve) created particular difficulties. In different contexts it could be translated as performance, capacity, capability, competitiveness, efficacy, efficiency, productivity, or proficiency. We eventually chose to translate this chameleon word as "performance" although this does not preserve the etymological link to achievement that is present in the German.

The English adjective "special" (etymologically related to species) originally referred to something that was specific or particular to an individual person or thing. Nineteenth-century German science made a distinction between *speciell* (special) and *allgemein* (general) treatments of a subject. Roughly speaking, *speciell* studies were detailed empirical investigations of particular cases whereas *allgemein* presentations were theoretical statements of general principles. We translate *speciell* as "special" in "special work," "special study," "special investigation," and "special communication" but note that "special" is intended in this distinctive sense of detailed studies of specific cases.

DER KAMPF DER THEILE
IM ORGANISMUS.

EIN BEITRAG ZUR VERVOLLSTÄNDIGUNG DER MECHANISCHEN
ZWECKMÄSSIGKEITSLEHRE

VON

DR· WILHELM ROUX,

PRIVATDOCENT UND ASSISTENT AM ANATOMISCHEN INSTITUT ZU BRESLAU.

LEIPZIG,

VERLAG VON WILHELM ENGELMANN.

1881.

THE STRUGGLE OF PARTS
WITHIN THE ORGANISM

A CONTRIBUTION TO THE COMPLETION OF THE MECHANICAL
THEORY OF PURPOSIVENESS.

BY

DR. WILHELM ROUX

PRIVATE DOCENT AND ASSISTANT AT THE ANATOMICAL INSTITUTE OF BRESLAU.

LEIPZIG,

WILHELM ENGELMANN PUBLISHERS.

1881.

Foreword to the First Edition (1881)

Many authors have already recognized to a greater or lesser extent that the new doctrine of the development of the organic kingdom as it was created by its founders, despite many eminent achievements, is not quite sufficient to derive all arrangements of the organism. Depending on whether or not these authors view this entire field favorably, the shortcomings, ignoring all achievements, are sometimes emphasized in the most exaggerated manner; sometimes weighed up by calm judgment; and sometimes the authors barely dare to even hint at them softly. Despite this multifaceted criticism and the studious work to fill the gaps, there seems to be much that remains to be completed and still many shortcomings to be revealed.

If, in the following, I emphasize the incompleteness in one of the least noticed directions, in the direction of the emergence of the finer inner purposiveness of the organization of animals, and attempt to supplement it, it is not done with the intention of diminishing the work of those great men, which incidentally would hardly be possible, but is done to round off and strengthen the theory according to the measure of my powers.

If the essentials of what is to be achieved by the introduced principle have been emphasized as far as possible compared to what is already known, this was not done to puff up the author at the expense of their merit but to make the introductory statement as clear as possible by emphasizing its peculiarity. But once what is new has been recognized and received, it will, of its own accord and commensurate with its importance, be apprehended and integrated as a modest part of that whole from which it has sprung and which has supported it and which it in turn wishes to support.

First of all, however, what is submitted here to the public is a mere sketch of the topic covered. The factual grounds of what is deductively developed may sometimes appear somewhat hypothetical in detail, and perhaps it cannot be otherwise when a new idea is founded largely on old observational material acquired with earlier principles. Nevertheless, I sustain the hope that the more recent observations that I cite, which form the foundation, can support the conclusions, and that the ideas of the "trophic effect of the functional stimuli" and the "principles of direct functional self-organization of so-called purposiveness" that follow from it will be acknowledged, whether now or only after further support has been gained through the work of the next few years. In order to link the new with the current views more easily, I did not neglect to point out that very similar principles of trophic stimulation already enjoy a more or less justified recognition in the teachings on trophic nerves, the development of tumors, and sensory perception (according to Ewald Hering). However, I did not base myself on these grounds in either the conception of my ideas or the present writing.

The low esteem for theoretical deductions in some circles (which is far below the esteem awarded the most insubstantial objective description) prompted me not to spend too much time on the execution and presentation of this first draft on the one hand, and to avoid exhausting the reader with too much detail on the other. But I believe that the comprehensibility and certainty of the whole has not been significantly impaired by this and that, for the reasons given, one will be happy to ignore the shortage or brevity of the historical presentation and be satisfied to find citations only where it is to provide a direct confirmation.

And so I commend this work to scientific attention and criticism.

W. Roux
Breslau, October 1880

9. Entwic
Rhizomor
Sklerotien

Nr. 4.

Der

ZÜCHTENDE KAMPF DER THEILE

oder die

„Theilauslese" im Organismus.[*]

Zugleich eine

Theorie der „functionellen Anpassung".

Ein Beitrag zur Vervollständigung der Lehre von der mechanischen Entstehung des sogenannten „Zweckmässigen".

Von

Dr. Wilhelm Roux,

Privatdocent an der Universität und Assistent am anatomischen Institut zu Breslau.

Πόλεμος πατὴρ πάντων.
Heraklit.

Leipzig.

Verlag von Wilhelm Engelmann.

1881.

[*] Der ursprüngliche Titel lautete blos: „Der Kampf der Theile im Organismus".

The

CULTIVATING STRUGGLE OF PARTS

or the

"Selection of Parts" in the Organism.[1]

At the Same Time This Is a

Theory of "Functional Adaptation"

A Contribution to the Completion of the Understanding of the Mechanical Genesi
of the So-Called "Purposive."

By

Dr. Wilhelm Roux

Private Docent at the University of Breslau, and Assistant at its Anatomical Institute.

War is the father of all things.
Heraclitus.

Leipzig.

Wilhelm Engelmann Publishers.

1881.[2]

---

1 The original title was merely: "The Struggle of Parts in the Organism".
2 The "second edition" was published in 1895 but appears with the date 1881 on its title page. —Trans.

Foreword to the Second Edition (1895)

I t seems appropriate to preface this second printing of *On the Struggle of Parts in the Organism* with a few introductory words and a brief overview of its reception as well as of the effects of the same.

Because this treatise presents a youthful theoretical work, written when the author had only recently emerged from the scientific school at Jena of the first quinquennium of the seventies (from the lessons of Ernst Haeckel, Karl Gegenbaur, and William Preyer), it is natural that the teachings of these researchers form the basis of its content. I must confess that the latter was the case to such an extent that I wrote the text largely unconsciously; because when I put down the stenographic pen, I usually did not know what the pen had just written.

So my brain worked with this material like a machine that has been tuned for a particular task. My deductions and the drawing of conclusions from previously absorbed fundamental ideas took place of their own accord. Thus, the fundamental ideas I had absorbed were spared from any critical examination. Such criticism followed only later, partly by myself independently, and partly suggested and spurred by other authors.

Thus, my views have changed significantly on some fundamental questions, for example, with regard to Haeckel's tenets regarding the homogeneity or structurelessness of protoplasm, which at the time was considered indubitable, and the soundness of the evidence for the "bequeathal of properties acquired by the individual," among others. Even today the latter question cannot be regarded as definitively decided in the negative sense; and this uncertainty makes itself felt in the most disruptive way in all causal phylogenetic inferences.

Because I believed I was not permitted to make any significant changes to the original text which has priority to my changes of mind, I have limited myself to using square brackets to add my current thoughts as comments. These square brackets also contain new evidence and discussions (or refutations) of new objections.

Regarding the presentation, on several occasions friends, especially Professor Gustav Born in Breslau, made me aware that it was often difficult to understand. Sometimes this was partly due to formal shortcomings. Sometimes it was because my chain of reasoning mentioned only the beginning and end of the chain. I found these objections to be justified only too well when I read through the text again, and I could not bring myself to print a second edition still containing these shortcomings. These defects were therefore eliminated as far as possible, and the diction has been improved and made more precise in many places. I hope that these improvements have made the previously difficult to understand text quite easy to read, and that these improvements together with the expression of my current views may win new readers and perhaps another reading from previous readers.

Now let us look at the reception this treatise has found so far in the scientific literature.

Charles Darwin described it in an 1881 letter to George Romanes as "the most significant book on evolution that has appeared in some time." He adds, "I think it is a huge oversight of Dr. Roux that he never considers plants; they would simplify his problem."[1] Finally, Darwin asks the letter's recipient to write a review of my book for *Nature*.

---

1 [This reproach is certainly justified. It would be desirable for someone competent to fill the gap. The plant physiologist Carl von Nägeli's great work, *The Mechanophysiological Theory of the Ancestry-Doctrine,* cannot be considered as such a complement; this author does not know my work, nor does he independently make use of the struggle of the parts. The same is true of Gustav Tornier's semi-botanical *The Struggle for Food.*]

Romanes complied with this request, but only in a very incomplete manner, for he merely provided a very general presentation of the goal of the book, emphasizing in particular that Herbert Spencer also drew analogies between physiological and social organisms, which included the struggle for existence, and that Spencer also publicized what I termed "functional adaptation" but which Spencer called "direct equilibration."

If Romanes had gone into a little detail, he would have been able to report that my book specifically rejected a theory long accepted by Spencer, and which was already widely known in Germany before Spencer, the explanation of functional adaptation by functional hyperemia, and that I subsequently offered a new explanation, on the basis of the struggle of equal parts in the organism, of the tissue qualities bred by that struggle, which sufficed even for the most delicate adaptations that were only recently discovered.

The fact that Mr. Romanes left the essentials of my book completely unmentioned and, above all, emphasized the agreement of its final goal with some of Herbert Spencer's writings is probably the reason it has remained virtually unknown in England,[2,3] so Spencer and Alfred Russel Wallace do not seem to know of it. Spencer and Romanes, as far as I know, never cite my book, and Wallace only thinks of it in his presentation of the doctrine of descent on a special occasion and only after a passing quotation from August Weismann. The book found wider circulation and recognition in Germany.

In his famous work *The Natural History of Creation,* Ernst Haeckel wrote, "Closely connected to cumulative (or 'heaped-up') adaptation, and partly falling under it, are the important variations of organization, which have just recently been termed 'functional adaptations' and explained

---

2   The effect of such a review by a man so respected in England in the most widely read English journal, passing over the essential content of the book in silence, was so intense and so powerful a rejection that even the Duke of Argyll's appended discussion failed to generate any interest in the original text.

3   The back-and-forth correspondence in *Nature* during October and November 1881 between Romanes, the Duke of Argyll, and William Carpenter is entitled "The Struggle of Parts in the Organism" but exclusively addresses Romanes's dismissal of the argument by design in the first paragraph of his review. None of this correspondence mentions Roux's book. —Trans.

clearly and in great detail by Wilhelm Roux. His treatise on *The Struggle of the Parts in the Organism* (1881) is one of the most important new contributions of the substantial Darwinian literature."

After discussing the struggle for existence among individuals, Haeckel also says, "No less important however, indeed fundamentally of much greater and general significance is the struggle for existence that takes place everywhere and at all times among all the constituent parts of which these individual beings are formed. Their transformations are indeed only the total outcome of the particular development of all their constituent parts.

"Darwin himself did not go into detail about these elementary structural changes. Professor Wilhelm Roux in Breslau gave the first comprehensive presentation and critical illumination of the same in 1881 in his excellent work *The Struggle of Parts in the Organism, a Contribution toward the Completion of the Mechanical Theory of Purposiveness.* I consider this treatise to be one of the most important contributions to the theory of development to have appeared since Darwin's major work was published and to be one of the most important additions to selection theory."

The *Berlin Clinical Weekly* of 1882 has a review, signed Lds, which states, "An exquisite work, full of fertile ideas and with the endeavor to incorporate cellular physiology into the theory of development. . . . We can earnestly recommend reading this excellent mature booklet containing a kind of philosophy of contemporary morphology."

In the *Zoological Annual Report* for 1881, William Marshall in Leipzig says of the present work, "A curious book of great importance, which stands in a similar relation to the theory of descent as Virchow's *Cellular Pathology* to pathology!" Haeckel expresses the same idea in these words: "Accordingly, the selection of cells, as it occurs everywhere in the tissues according to Roux, could also be called *cellular selection,* in juxtaposition to the *personal selection* among independent individuals first demonstrated by Darwin. Cellular selection would relate to personal selection as Virchow's cellular pathology relates to personal pathology or, as set out by me, cellular psychology relates to personal psychology."

In a review entitled "A New Extension of Darwinism," the book was discussed in depth by Eduard von Hartmann, who showed complete mastery of its content and included his personal thoughts on the subject. Ernst Krause also reviewed the book in *Kosmos.*

The book's contents are recognized in Hugo Spitzer's *Contributions to the Theory of Descent and the Methodology of Natural Science,* written with rare acumen and a wealth of knowledge; also by Alois Riehl in his *Philosophical Criticism and Its Meaning for Positive Science;* in Wilhelm von Reichenau's essay "On the Origin of Secondary Male Sexual Characters, Especially in the Leaf-horn Chafer"; by August Weismann in his works *On heredity* and *The Germ-Plasm;* in Carl Claus's essay *The Value of Natural Selection as an Explanatory Principle;* in Gustav Guldberg's presentation *On Darwinism and Its Scope;* in the great work of Julius Wolff, *On the Law of the Transformation of Bones* for the explanation of the direct adaptation of bone-structure; by Élie Metchnikoff for the explanation of immunity against infection; further by Hermann Nothnagel in his keynote address to the second general session of the International Medical Congress in Rome on "The Adaptation of the Organism to Pathological Changes," and by others.

Along with these appreciative judgments there are also derogatory judgments which I must not withhold from the reader.

Friedrich Merkel characterizes the book's contents with the words: "Roux illuminates the well-known fact that only a balance of the complete functions of the individual, be it of the whole organism or of the individual cells, guarantees the *status quo,* using the technical terms of the doctrine of descent."

In contrast to the above philosophers, Wilhelm Wundt recognized that in this book "such life phenomena that one had previously understood in terms of their causal relationships" are "reinterpreted in teleological form" (1881, 437). Wundt did not however respond to my consequent request as to who had discussed before me the manifestations of life presented in my book "in their causal relations," or of what my "teleological reinterpretation" (223) of the same might consist.

Heinrich Spitta of Tübingen seems to have penetrated the content of the book just as deeply as these authors; because, apart from the content of the preface, he has only taken from it the fact that it consists of six chapters, and that on the second page, in a sentence that attracts attention because it is in blocked printing, a serious error is found, instead of "so-called purposiveness" simply appears "purposiveness." "So-called" is printed a few lines before, but not blocked; and from the rest of the content it would have been clear that the term "enhancement of lasting-ness" is introduced as a replacement for the unwanted purposiveness.

The anonymous reviewer of the *Central Literary Review* apparently read the book, but probably not with complete understanding; for he reports that the book transfers the struggle for existence "to the chemical molecules" (instead of to the ultimate living molecules capable of assimilation and multiplication, the "ultimate organic process units").

If we now take a look at the influence that the book has had, it mainly concerns the *doctrine of descent* and its philosophical evaluation, as can already be seen from the above indications.

Further consequences are drawn from the principles set out here in Georg Pfeffer's interesting work "The Transformation of Species, a Process of Functional Self-organization."

In the case of Hans Strasser, the idea of the selection of parts fell upon very fertile ground because he skillfully exploited this theory in his thorough work, *On the Knowledge of the Functional Adaptation of the Striated Muscles*. Paul Fraisse's *The Regeneration of Tissues and Organs in Vertebrates* and Dietrich Barfurth's "On the Regeneration of Tissues" similarly use the idea to explain processes during regeneration. The latter author also made a valuable experimental contribution to functional adaptation.

Because the book deals with the finest processes of selection and formation among the parts of the organism, it uses a large number of less well-known anatomical and physiological facts, and this seems to have restricted its dissemination considerably.

As a result of the former circumstance, one could have expected that *physiology* would have taken up the concept of selection of parts and used it in specific empirical pursuits. However, this did not happen, given the current limitation of physiological research to the mere "operation" of the animal machine that is accepted as "given." Only Emil Du Bois-Reymond has taken on one part of the subject under discussion. In his academic lecture "On Exercise," given six months after the publication of this book, he treated functional adaptation in the same way as I did and also used the explanatory principle that I first established and which was thoroughly established in chapter III. (However, since he accidentally only thought of my contribution regarding a very particular, narrow matter, Du Bois-

Reymond has since expressed (by letter) the intention to amend this when his address is next reprinted.)

Pathology should have been particularly strongly motivated to seize the idea of the selection of parts in the organism because with every disease, be it chronic or acute, general or local, as a result of the change in the living conditions of the cells an opportunity arises to eradicate the nonresistant parts while leaving the resistant parts, whereby a corresponding internal recultivation, systemic or local, must result if the individual survives the disease; so, for example, the organism becomes an economizing machine through chronic inanition or becomes more resistant to a poison by surviving poisoning, and by this means becomes habituated to poisons and medicines and *immune* to infectious germs.

These ideas would be worth examining pathoanatomically and pathochemically but, despite much recent work on immunity, were only considered theoretically by three authors, Paul Grawitz, Gustav Wolff, and especially by the excellent zoologist and pathologist Élie Metchnikoff, the former authors without reference to the present work or my other work that preceded it in 1879.

Theodor Ackermann's rectoral address, *Mechanism and Darwinism in Pathology,* expresses a considerable correspondence of ideas and idioms with the relevant parts of this treatise; but he missed the importance of the most important idea for pathology, the internal recultivation and the possibility, based on it, of habituation to harmfulness and immunity.[4]

Orthopedics in particular would also have reason to seize the content of this book because it contains the basis for a scientific orthopedic procedure. Raw empiricism can gradually be turned into action based on an understanding of the processes, as I have repeatedly but unsuccessfully emphasized, only if one knows the "formative reactions" of each of the tissues involved in the deformity of a part of the body, such as when the

---

4 [In a recent publication from 1894, "The Pathological Formation of New Connective Tissue in the Liver and Pflüger's Teleological Causal Law," Ackermann says that I had "only made more specific" Pflüger's law. It had thus escaped Ackermann that the essential content of my booklet culminates in causally deriving this teleological law, ingeniously formulated by Pflüger as factual, by explaining it mechanically and thus undressing its apparently metaphysical character.]

spine is curved, the reaction of the bone, cartilage, and connective tissue that composes it, and, moreover, of the muscle tissue that influences it.

The doctrine of "functional adaptation" is the scientific basis of orthopedics because the latter must first and foremost be "functional orthopedics." Therefore, orthopedists have the most practical interest in maintaining this doctrine; and they themselves should compete with causal morphologists to increase our insight through the indicated analytical experiments. Up until now, however, only Julius Wolff has done this but restricted himself exclusively to bone tissue.

The following is the overall result of the spread and evaluation of the principles set out in this book:

A thought that is new and alien to the usual view spreads only slowly. A book that, in its title or the beginning of its text, reveals that its content goes beyond the boundaries of one of the special fields into which biology is currently divided is, for this reason, almost never read by any representative of one of these fields. Few researchers today still have the universal striving of biologists in the middle of this century to gain an overview of life operations as a whole.

CHAPTER I

Functional Adaptation

A. Its Achievements

The problem of explaining purposiveness in nature already occupied the oldest philosophers in the classical age of antiquity and through Empedocles found its general and, in principle, complete solution. He already achieved the ultimate goal of a doctrine of purposiveness: the realization of how it could arise mechanically in the absence of a formative force with deliberate goals.

This great thinker understood the basic material substance to be the changeless primal existence and let it be mixed and shaped by forces of love and hate. In this mixture of materials, provided with two opposing forces, a long-lasting back-and-forth struggle must take place, out of which only the "lasting" aggregations would ultimately remain alone because every grouping would continue to be broken apart, again and again, until stronger conglomerates were formed by the back-and-forth action.

Empedocles thus first discovered the possibility that so-called purposive arrangements could come into existence purely mechanically through elimination of all nonlastable combinations by the back-and-forth actions of forces. And so it was demonstrated how a wonderfully purposive animal could have a mechanical origin (at least philosophically).

This so-called purposiveness was not intended but came into being, not teleological but natural-historical, emerging mechanically over a long period. For what persisted did not correspond to some preconceived purpose but possessed properties necessary for "existence" under the given circumstances. [That is why I have continued to use the word "lastability"[1] in its place.] This is the only sense in which I will use the word "purposive" in what follows.

One might have thought that this philosophical solution of the problem would soon have been followed by an empirical solution. But whoever knows the history of Greek philosophy knows how tightly the Greeks were bound by their worldview, partly by lack of empirical observation, partly by false observations that led to deluded beliefs, and partly because the ability to observe truly objectively and self-critically was reserved for only a few of the most important men.

Thus, the significance of the Empedoclean solution to this great problem was not realized, let alone that it could be used for special research. It was lost completely and had to be discovered anew on the basis of burdensome, detailed, empirical research, after long and fruitless searching by many excellent men. At least this time it was not just an in-principle philosophical solution but an exact scientific solution.

As is well-known, Charles Darwin and Alfred Wallace not only rediscovered the principle of struggle as the cause of the mechanical origin of the purposive, but at the same time demonstrated that, because of the geometrical multiplication of organisms, such a struggle must take place among them, with an ever-present possibility of improvement as a result of the constant variation of organisms in all their parts.

Because the left-behind beings bequeath their privileged properties together with coincident new modifications built on this foundation, the possibility is given to select the best of these new modifications—on average, already more perfect—so that a constantly increasing perfection must take place. At the same time, this process of perfection will lead to a multiplicity of forms if, as is actually the case, the external selective conditions themselves are diverse and change over time.

---

1 This is Roux's only use of *Dauerhaftigkeit. Dauerfähigkeit* is his usual term. —Trans.

With the demonstration of the effectiveness of the principle of selection and the auxiliary principles of the variability of organisms and of the external conditions of existence, the necessity of the emergence of a steadily enhanced multiplicity and adaptedness to the *external* conditions was demonstrated. At the same time the possibility was opened up that the great diversity and complexity of the higher organisms could have been derived by gradual transformation from lower simple, even simplest, states. The doctrine was developed for the purpose of demonstrating the origin of species through gradual further differentiations in certain directions and the descent of higher from lower organisms, and it has been worked on diligently for twenty years toward completion of this end.

On the other hand, the origin and causes of purposive *internal* arrangements were less investigated, both for those arrangements that represent specific characters and especially for the more general characters shared by entire classes or orders. Therefore, the doctrine has not yet been thoroughly examined in detail as to whether it is capable of revealing all existing internal purposivenesses of organization as necessary consequences of mechanical principles already established or whether other auxiliary principles must also be accepted and demonstrated.

After careful examination, I have come to the latter point of view, which I intend to present here. For this end, I must demonstrate both that existing principles are insufficient and that one or more other principles have been active. I came to this conclusion when attempting to use the existing doctrine of descent to explain the individual purposivenesses found in organisms; and in the following I want to stick closely to this task, and therefore I will accept the doctrine of descent as already fully established and sufficiently known to readers insofar as it concerns other relations.

According to the accepted doctrine of descent, the purposive arises predominantly or almost exclusively through the selection of random formative variants: first in the struggle for existence and second through sexual selection. The first of these two principles is purely mechanical, but no definitive judgment can yet be made with regard to sexual selection because of its dependence on mental influences. Because sexual selection is almost completely irrelevant for our purposes, we can open a common selection account (*Ausleseconto*) for natural and sexual selection when investigating their performance.

1. Purposive Effects of Increased and Decreased Use

In addition to derivation of the purposive from natural and sexual selection among individuals, the founders of the doctrine of descent already identified a principle of transformation that provided a shorter path than selection among favorable variations to bring about purposiveness within the individual. This was the Lamarckian principle of the effects of use and disuse, which different authors have admitted as contributory to very unequal degrees; this is partly because the heritability of these effects is obscure and partly because one knows nothing of its causes and therefore cannot tell whether it is a mechanical principle, possibly to be used, or a metaphysical or teleological principle to be used as little as possible.

Detailed investigations into the heritability, cause, and mode of action of this principle have been lacking. In the following, we intend to contribute something to the completion of knowledge in these three directions. The investigation of the mode of action will lead us to those purposive arrangements that cannot be derived directly from the aforementioned mechanical principles of selection of Darwin and Wallace, nor can the effects of use and disuse be derived from these principles alone.

Darwin expresses himself as follows with respect to the effects of use and disuse (which we shall subsume under a somewhat more general concept, briefly *functional adaptation*,[2] to be discussed below):

"Changed habits produce an inherited effect, as in the period of flowering with plants when transported from one climate to another. With animals the increased use or disuse of parts has had a more marked influence; thus I find in the domestic duck that the bones of the wing weigh less and the bones of the leg more, in proportion to the whole skeleton, than do the same bones in the wild-duck; and I presume that this change may be safely attributed to the domestic duck flying much less, and walking

---

2 [I understand "functional adaptation" to mean adaptation to a function by its exercise. This adaptation extends to the size, shape, structure, and quality of the organs. The word "adaptation" is used in the sense usually understood. According to this, "adaptation" of living beings to any circumstances is a change in living beings that makes their lastingness greater than it would be under the same circumstances without this change. What matters is the enhancement of lastingness.]

more, than its wild parent. The great and inherited development of the udders in cows and goats in countries where they are habitually milked, in comparison with these organs in other countries, is another instance of the effect of use."[3]

He adds, "Some (perhaps a great) effect may be attributed to the increased use or disuse of parts." The words in parentheses that increase the emphasis are not found in the first edition of the book. And on a later page he adds, "From the facts alluded to in the first chapter, I think there can be no doubt that use in our domestic animals has strengthened and enlarged certain parts, and disuse diminished them; and that such modifications are inherited. Under free nature, we have no standard of comparison, by which to judge of the effects of long-continued use or disuse, for we know not the parent-forms; but many animals possess structures which can be explained by the effects of disuse." He cites the American logger-headed duck that can only flap along the surface of water, the inability of the ostrich to fly, and the stunted front tarsi of many male dung-beetles.

And he also remarks, "The eyes of moles and of some burrowing rodents are rudimentary in size, and in some cases are quite covered by skin and fur. . . . It is well known that several animals, belonging to the most different classes, which inhabit the caves of Carniola and of Kentucky, are blind. In some of the crabs the foot-stalk for the eye remains, though the eye is gone;—the stand for the telescope is there, though the telescope with its glasses has been lost. As it is difficult to imagine that eyes, though useless, could be in any way injurious to animals living in darkness, their loss may be attributed to disuse."

After citing the example of a barnacle that, more or less, loses its own calcareous shell when it lives on another as a freeloader, Darwin immediately weakened the significance of the principle he had just admitted by noting "thus, as I believe, natural selection will tend in the long run to reduce any part of the organisation, as soon as it becomes, through changed habits, superfluous, without by any means causing some other part to be

---

3 Roux quotes from the fifth edition of *Über die Entstehung der Arten* which corresponds to the sixth edition of Charles Darwin's *The Origin of Species*). We quote directly from Darwin's sixth edition. —Trans.

largely developed in a corresponding degree. And, conversely, that natural selection may perfectly well succeed in largely developing an organ without requiring as a necessary compensation the reduction of some adjoining part."

From this, but also as the conclusion of his entire work *On the Origin of Species,* Darwin ascribes only a small role to the direct transformative effects of use and disuse despite his recognition of the principle, and he derives most of the diminution of unnecessary organs and enlargement of useful organs from the action of selection from free variations. However, his example of the reduction in size of the calcareous shell, which could not have resulted from gradual subsequent atrophy, seems to speak against his conclusion.

Haeckel ascribes much greater importance to effects of use and disuse. He derives them from nutrition and proves (without going into a more precise explanation of their direct effect of shaping the purposive) that these changes in habit are ultimately only caused by changes in external circumstances, and then explains in detail how large are the resulting changes. The muscles of gymnasts double in thickness, and their powers of performance are increased many times. He says, "The act of exercise it-self, the often repeated movement of the muscle, causes a change in the nutrition of the muscle, which causes greater influx of nutrients. The muscle thereby grows, increases the number of its primitive fibers, perhaps also those chemical constituents of the muscle substance which are primarily active in its contraction; it therefore probably improves not only quantitatively, but also qualitatively, as the fat deposited in the unexercised muscle disappears through exercise and is replaced by more noble protein constituents."

He also states, "How powerfully this law of habituation acts is so well known that we do not need to provide any more examples and only need to recall the well-known proverb: *Consuetudo altera natura.*[4] We want to emphasize that the disuse of organs, which has a regressive effect on them that is so significant for dysteleology, is no less important than the use of organs."

---

4 Habit is other nature. —Trans.

But his greater estimate of the importance of functional adaptation rests chiefly on the high heritability that he ascribes to its formations. His "law of adapted or acquired bequeathal" claims in general, "Under favorable circumstances, the organism can bequeath to its offspring all characters that it acquires through adaptation during its individual existence that its ancestors did not possess." He adds, "Much more important than the monstrous, conspicuously protruding changes, which are transmitted by adaptive bequeathal, are the inconspicuous and minor changes, which only acquire their great importance for the remodeling of organic forms through accumulation and strengthening over the course of generations." He further says that bequeathal is all the more certain, and occurs more fully for all following generations, the more sustained the action of the conditions of causal adaptation and the longer they continue to act on subsequent generations.

From the outset, Haeckel therefore deviated significantly from Darwin, who, despite the selected appreciative examples in his first work *On the Origin of Species,* did not consider acquired characters to be sufficiently heritable to give them a significant influence on the effect of selection. That Darwin has not changed this opinion in his most-read work, even in its most recent editions, is probably the reason that his actual change of opinion, explained in detail in *On the Variation of Animals and Plants under Domestication,* has not been sufficiently appreciated. Consequently, some of his supposedly most devout followers, Georg Seidlitz for example, accuse other thinkers who ascribe greater importance and heritability to functional adaptation, such as Haeckel and Oscar Schmidt (indeed Darwin himself), of apostasy from the supposedly true doctrine.

As we shall see in a moment, Darwin (in *On the Variation of Animals and Plants under Domestication*) subscribes almost entirely to the views expressed by Haeckel (in his *General Morphology*). In his summary of hereditary variability arising from use, Darwin says, "Increased use adds to the size of a muscle, together with the blood-vessels, nerves, ligaments, the crests of bone to which these are attached, the whole bone and other connected bones. So it is with various glands. Increased functional activity strengthens the sense-organs. Increased and intermittent pressure thickens the epidermis; and a change in the nature of the food sometimes modifies the coats of the stomach, and increases or decreases the length of the intestines. Continued disuse, on the other hand, weakens and diminishes all parts of the organisation. Animals which during many generations have taken but little exercise, have their lungs reduced in size, and as a consequence the bony fabric of the chest, and the whole form of the body, become modified. With our anciently domesticated birds, the wings have been little used, and they are slightly reduced; with their decrease, the crest of the sternum, the scapulæ, coracoids, and furcula, have all been reduced." However, he greatly limits the effects of disuse by saying, "With domesticated animals, the reduction of a part from disuse is never carried so far that a mere rudiment is left, but we have good reason to believe that this has often occurred under nature. The cause of this difference probably is that with domestic animals not only sufficient time has not been granted for so profound a change, but that, from not being exposed to a severe struggle for life, the principle of the economy of organisation does not come into action."

He also remarks, "Corporeal, periodical, and mental habits, though the latter have been almost passed over in this work, become changed under domestication, and the changes are often inherited. Such changed habits in any organic being, especially when living a free life, would often lead to the augmented or diminished use of various organs, and consequently to their modification. From long-continued habit, and more especially from the occasional birth of individuals with a slightly different constitution, domestic animals and cultivated plants become to a certain extent acclimatised, or adapted to a climate different from that proper to the parent-species."

Darwin therefore grants, in this work, a much more considerable influence to the effects of functional adaptation on the transformation of organisms, alongside those of natural selection, than he does in the *Origin of Species.* Since the changes brought about by functional adaptation are directly purposive, Darwin thereby recognizes a principle that directly produces what is purposive by a much shorter path than by selection. He thus places these two principles into strongest competition and gives the impression of wanting to reintroduce a dualism that had been happily considered to have been eliminated.

Alfred Wilhelm Volkmann has already said, "Selection is not enough to explain the mutual dependence of organs." He recalls Georges Cuvier's saying that one only need examine the mandibular joint of a mammal to determine whether one has the bones of a carnivore, ruminant, or rodent.

In terms of its occurrence, the extent of the effects of frequent use on the individual organs is completely exhausted by Darwin's examples: for he shows effects on all organs, or assumes them for the sense organs for which he has not proven a direct transformation or strengthening by functions. However, in the latter case we are unable to distinguish whether the sense organs themselves have become sharper or whether the ability to perceive the stimuli supplied by the sense organs has merely improved in the brain, or whether exercise affects the end-organs themselves or just the central organ. The only relevant anatomical observation comes from Johann Gudden who found that in newborns the olfactory bulbs enlarged beyond the usual size when both eyes of the animals concerned were extirpated and the ears were closed. But this fact only indicates a change in the central organs and does not exclude the possibility of change in the end-organs.

For adaptation of the central nervous system to certain modes of use, I would like to cite an apt example provided by Hermann von Helmholtz who says, "If one places prisms with a refracting angle of 16 to 18 degrees in front of both eyes so that both prisms cover the external objects, then move to the right, and look at any object for its precise position, then close your eyes and reach for it, then of course you reach past it on the right. But if one manipulates things using these glasses for only a few minutes, one will certainly grasp the object in the case of repetition. In this short time, the whole combination of innervations in the extremities has changed and adapted to the new experiences. If one now takes the glasses away, one reaches past objects on the left, because the new type of innervation no longer fits the old conditions."[5]

Sigmund Exner remarks very aptly, "It is also necessary that our combinations of innervations are highly modifiable. If this were not the case,

---

5   The "quotation" from Helmholtz is copied from page 249 of Exner (1879). Exner appears to be paraphrasing Helmholtz. Many of Roux's citations appear to be similarly at second-hand (Exner and Roux cite the same page of Helmholtz's *Physiologische Optik*). In Exner (1879), the paraphrase of Helmholtz is immediately followed by the words that Roux quotes from Exner in Roux's next paragraph. —Trans.

we would lose the ability to carry out correct combinations of movements when the muscular apparatus was fatigued, and even more when individual muscles were unevenly fatigued." Thus, the ability to adapt functionally is a prerequisite for the acquisition of any physical dexterity, and exercise is nothing more than the development of such adaptations in the organism; nay, the fixation of all sensory impressions in the cerebral cortex must be understood as a direct functional adaptation to the outside world.

Furthermore, a peculiar behavior can be cited here. According to Jean-Marie Philipeaux, Alfred Vulpian, Elias Cyon, Moritz Schiff, Rudolf Heidenhain, and some of Ludimar Hermann's students, after cutting through the motor nerve of the tongue (*nervus hypoglossus*), a taste nerve of the tongue (the chorda of the *nervus facialis*) acquires a motor effect on the tongue such that the tongue now rises on irritation of the chorda, an effect that disappears after restoration of the motor nerve. Temporary vicariation (substitution) of nerves is certainly a striking degree of functional adaptation.

Acceptance of the direct adaptation of bones to new conditions meets, in my experience, with particular resistance from those who have not yet observed it themselves. It therefore does not seem superfluous to mention a particularly demonstrative case from my own observations. It concerns an anatomical preparation of an unhealed fracture of a tibia broken in the middle. The two broken ends of the tibia were rounded and thinned, whereas the fibula was thickened in its entirety to six to eight times the normal cross section while maintaining approximately the normal shape and especially a perfectly normal smooth surface that excluded inflammatory bone formation. The heads of the fibula were less thickened, but shaped in such a way that they were able to fulfill the new functions of transferring pressure from the upper end of the tibia to the lower end by means of very strong strands of connective tissue between the heads of the fibula and each associated end of the tibia. Such examples of activity hypertrophy of the bones and connective tissue will probably be found in every pathological collection.[6]

6 This preparation was the subject of a detailed analysis by Roux (1885) and is featured prominently in Julius Wolff's *Das Gesetz der Transformation der Knochen* (1892). —Trans.

Eduard Pflüger mentions quite generally, "But it is a fact that with greater loss as a result of increased work such conditions arise, according to which something more is always regained than was lost, for the continued greater use of the organ causes its mass and power to increase"

With the extension of the reshaping effect of functional adaptation to all organs, it is implicitly and also explicitly stated that all tissues of the body are affected, including ganglion cells, nerves, sensory cells, muscular, glandular, epithelial, connective, cartilage, and bone tissues. But so little has the nature of the effect been taken into account that Darwin and all the other authors merely mention that increased use increases the size of the organs while decreased use makes them smaller.

It seems to me, however, worthwhile to examine the mode of action on individual organs. The following *morphological law of functional adaptation* emerges with a high degree of probability simply by examining what is currently known without making any new special observations: *With intensified activity, each organ enlarges only in that dimension, or those dimensions, which strengthen the activity.*

This "law of dimensional activity-hypertrophy" manifests itself most clearly in the behavior of muscles that have been enlarged by greater functional use. While the muscle gradually thickens, possibly up to twice its original cross section, its length remains unchanged; or it increases so minimally that no one has noticed, and there are reasons to expect, on the contrary, a shortening. Enlargement has therefore been restricted to the two cross-sectional dimensions.

Microscopic examination of such a muscle shows that its individual fibers are somewhat thicker than those of less active muscles of the same individual, but by no means to the extent that the thickening of the whole organ can be ascribed to this alone; rather, there has also been an increase in the number of fibers. According to Virchow, we should analytically distinguish the former phenomenon, hypertrophy (the enlargement of the specific elementary parts of the cells), from the latter phenomenon, hyperplasia (the increase in the number of specific elementary parts), even if both usually occur at the same time.

In the present case, the hypertrophy of the individual muscle fibers has been limited to the two cross-sectional dimensions without lengthening in the third dimension. The absence of the latter can be seen from external observation of short muscles, whose entire length is formed by only one fiber, and also of long muscles (the length of which is made up of several muscle fibers strung together). This is shown in the lack of an elongation of the whole organ which must necessarily have occurred if the elementary parts had lengthened—unless, that is, fibers change their position relative to others by pushing themselves together more in the direction of length, or some fibers elongate but others shorten accordingly in other parts of the muscle, both in themselves equally improbable processes, quite apart from the lack of fit to the behavior of the shorter muscles. But the fact that the muscles must strengthen their activity in cross section needs no explanation.

Why do the newly formed protoplasmic particles of the fiber arrange themselves only in the two cross-sectional dimensions but not in length? Why do the newly formed muscle fibers do the same?

Deviations from this typical behavior often occur in the heart and other cavity-enclosing muscles of the bladder, stomach, intestines, and uterus in that thickening is also associated with proportionate or nonproportionate lengthening of the fibers and thus enlargement of the enclosed cavity. It is precisely the fundamentally different behavior in these localities that gives us a significant clue as to the cause of the phenomenon observed above in skeletal muscles.

The behavior of the tendons and ligaments should be cited as further evidence of the law outlined above. As is well-known, these likewise do not lengthen but simply thicken with stronger function. Lengthening, if it were to occur, would immediately reduce the lifting function. So here, too, there is just an arrangement of new molecules and fibers in the cross-sectional dimensions.

Perhaps the unequal thickness of nerve fibers, as observed in every cross section of a nerve trunk or the spinal cord, is due to an unequal strength of function. Here again, there is no elongation because fibers are not particularly tortuous in nerve trunks. [This supposition does not contradict Gustav Schwalbe's interesting finding that longer nerve fibers are thicker, because longer fibers must be excited more strongly than shorter fibers in order to transmit an impulse of the same strength.]

Hypertrophy of acinous glands manifests itself both in an increase in the number of the glandular lobules and an enlargement of each lobule.

Because the glandular epithelium remains in one layer with this enlargement, its hyperplasia must occur only in the two dimensions of the secretory surface. In tubular glands, the newly formed cells are almost exclusively attached to one another in the direction of the gland's length while the tube's thickening is solely due to hypertrophy of its cells. Because, however, in the final stages of development these organs have grown only in these dimensions, one could say that their betterment has simply taken place according to bequeathed developmental laws, if one does not want to relate the origin of these laws to the present principle.

After loss of substance, the epidermis multiplies only in the two dimensions of its surface, and this multiplication persists only until epidermal cells again meet epidermal cells on all sides. (According to Carl Friedländer, when this cannot occur in fistulas, the epidermis grows out over the entire fistula canal.) Any other stimulus causes epidermal cells to multiply in the one dimension of thickness to the complete exclusion of the other two dimensions. If these stimuli had acted to increase the surface area, then, with passive involvement of the dermis, wrinkles would have arisen as observed in the intestinal tract. In the latter, however, the folds are not caused by an increase in the surface epithelium but rather by an increase in the glands, and the corresponding increase in area of the surface epithelium and the mucous membrane probably occurs only passively.

For this epithelial example, an objection could be made that should be taken into account, namely that the resistance of the underlying thick dermis to folding, due to stronger growth of the overlying epithelium, is probably too great. Adjudication of these objections can only be decided by detailed special investigations for each organ.

Loose connective tissue lengthens gradually when stretched and becomes hypertrophied in this single linear dimension. As is well-known, the same thing also occurs in tight connective tissue when tension is prolonged or excessive, whereas, as outlined above, connective tissues normally strengthen only in their cross-sectional dimensions if the tension is excitatory and takes place with appropriate intermissions.

The fovea centralis of the eye is the location of the clearest and most frequently used vision. Retinal cones of the fovea centralis are tallest in the direction of incident light and at the same time narrower than on lateral parts of the eye. It is perhaps to be assumed that the greater function of these parts is achieved by their greater length and that their lesser thickness is a consequence only of the greater tendency for the cells to multiply caused by the greater functional stimulus. It would not speak against this view if these differences were congenital because they could well be bequeathed, even if they had originally arisen through use.

The spleen and lymph glands achieve their function of the formation of blood cells evenly in all three dimensions. With increased function, these organs enlarge equally in all three dimensions (as far as the surrounding space permits in the case of the spleen). The fact that cellular proliferation is not restricted to particular directions is proved by the observations that cells in these organs are never arranged in rows, as otherwise must develop, and are arranged in the same manner in hyperplastic and unenlarged organs.

I will not cite further examples here, in particular not to mention the most interesting, unequal behavior of the vascular wall in the various dimensions because I intend to do the special study myself, which is needed to safeguard the above law, based on new observations thereupon. I shall then also point out the characteristic differences between activity hypertrophy and the hypertrophy that occurs in some organs as a result of increased blood supply.

Of course, what is typical of the law emerges most evidently in those organs where the various dimensions have different functions and hence are changed by different circumstances. At present we see that not all dimensions of the organs are uniformly enlarged by strengthened function, even where (as in the case of muscles and ligaments) space would allow it, but only those dimensions that determine the strength of the function. The relationship is such that in those organs whose "specific function" occurs in one dimension, such as muscles, tendons, glands, and nerves, the "magnitude of function" is determined by the other two dimensions. Conversely, in other organs (such as the epidermis, vessel wall, fascia, and perhaps also the cones of the retina) that perform their specific function in two dimensions, the strength of the function is determined by the third.

At this point I abstain from any discussion of the cause of the above law, which I have therefore only given in factual version, although an indication that the function itself is the cause of the enlargement of the dimensions that take care of the strength of the function would have been obvious.

Functional hypertrophy does not always bring about "isomorphic growth," which is equal enlargement in all diameters proportional to overall size, but it can also form morphologically novel characters by the possible limitation of enlargement to one or two dimensions. New characters arise from functional hypertrophy, and also from the uneven enlargement of the various organs with the same increase in function, but mostly from the unequal distribution of the hyperfunction over the various organs of the body.

If this principle gives the possibility of every conceivable change of form, this possibility is further facilitated and quantitatively supported by the opposing principle of the reduction in size as a result of the reduction in function by "inactivity atrophy." In connection with this principle, all possible magnitudes can now also regress backward to a point of complete disappearance. Inactivity atrophy is also limited in its effects to the dimensions of organs that implement the size of their function, so a law of dimensional inactivity atrophy must be established for this. Here, too, there are again differences in some organs from the simple atrophy that results from a reduction in the blood supply, and I reserve the right to make special investigations and demonstrations of this as well.

But in order for these two principles to give rise to *transformations*, or remodelings of form, *persisting urgent reasons for other uses are necessary*. Such transformations happen in animals only by embryonic variations of some parts, which then have an altering effect on the functions of others, or by changes in the external conditions, but in man also by a will that persistently acts in the same direction—for example, as the result of a choice of a profession. A persisting, urgent reason for particular alternative uses is an indispensable precondition for the transformative effects of functional adaptation, and it must continue uniformly for many generations if its changes are to become hereditary.

As well as this quantitative, shape-influencing effect of functional adaptation, there is an almost unnoticed "qualitative" effect of increased and decreased use, namely an increase or decrease in the specific performance of organs. This was first demonstrated by Wilhelm Henke and Franz

Knorz, who found that the same volume of muscle substance could accomplish 20 percent more in the right arm than in the left arm (in cadets who had very strongly trained the right arm by much fencing), [namely 7.4 kilograms on the left, almost 9 kilograms on the right, on average a little over 8 kilograms]. Haeckel said the same thing at the same time in the passage cited earlier but without specification of direct measurements.

Furthermore, the investigations of Ernst Tiegel showed an increase in the specific performance of a muscle within a single short physiological stimulus period by showing that, with the same stimulus, the lift height (shortening of the muscle) increased for a while before decreasing due to exhaustion.

Everyday experience seems to confirm the same thing for the central nervous system. Everyone knows of movements that have been laboriously learned through years of practice—for example, playing musical instruments—that once they have been initially trained can later be carried out easily such that they finally take place by themselves as fixed mechanisms with almost no conscious innervation. One cannot assume here that the fibers connecting the ganglion cells of the spinal cord would have become so many hundred times thicker in order to bring about the decrease in resistance in the tracts simply by increasing the cross section. It is more probable that the connecting tracts, at the same time as the increase in their cross section, have become qualitatively more conductive and that the ganglion cells produce relatively more impulse upon excitation.

In the same way the organs of our psychic activity become better performing by more frequent and intensive use, as we say, through practice.

Everything we learn physically and spiritually is a product of "functional adaptation": without it we would be unable to learn anything. And everyone knows how much quicker and easier it gradually becomes to learn, not just to carry out what has been learned, which indicates an increase in the specific performance of the whole system as a result of multifaceted use.

We are therefore justified in adding a physiological law of functional adaptation to the above morphological law of dimensional hypertrophy for the organs mentioned. *The specific performance of organs is enhanced by strengthened activity.* It goes without saying that this law, like all organic laws of performance, is valid only to a certain degree; nor does it deny that overexertion can weaken performance. We do not know whether this law applies to the glands or the passively functioning organs such as bones and ligaments.[7]

We have already stated that it is currently unresolved whether this law applies to the sense organs or whether the strength-enhancing exercise in apprehending and differentiating the sensory impressions is merely cerebral, because the sense organs are mostly affected in the same way from the outside by the impressions, but the faculty of comprehension fails to be strengthened because of insufficient attention to the impressions. However, a qualitative change is known for sense organs connected with an enhanced performance by the functional act, which is perhaps better understood as active exercise rather than as passively learned tolerance or habituation. Such is the circumstance that we gradually learn to distinguish different intensities of what were at first overwhelmingly strong sensory impressions. But here again it is impossible to separate how much of this exercise occurs centrally in the brain.

[Furthermore, this adaptation of actively functioning organs that increases their specific performance is opposed by the law: *if the degree of function is reduced for a prolonged time, the specific performance of an organ is reduced.* Perhaps these laws also apply to the passively functioning organs, to connective and supporting substances (connective tissue, cartilage, and bones). At least this is suggested by the degeneration of the articular cartilage of unused joints and perhaps also by the greater fragility of bones in old age.]

In the case of glands, the diminution of their performance when activity is reduced, as is often observed, speaks in favor of qualitative functional adaptation. Balthasar Luchsinger has recently found that a few days after cutting through the nerves whose stimulation causes secretion of sweat on the hind paw, stimulation of secretion by the gland cells is no longer possible even with pilocarpine, and he suspects with good reason that this is because the gland cells have lost their excitability. [This information was withdrawn by the author due to objections by other authors and their failure to replicate. However, the digestive glands are weakened by

---

7 We have rearranged text between this and the following paragraph to clarify the flow of Roux's argument. —Trans.

starvation with clear structural changes in the cells. Here, of course, the lack of food may itself play a direct role.]

Every doctor knows how prolonged inactivity decreases the excitability of nerves and muscles. The pathological anatomy of high-grade cases shows not only this decrease but also qualitative changes such as the presence of fat granules in the protoplasm.

Besides, we cannot withhold from the reader the strange finding of Christiaan Karel Hoffmann and Sigmund Exner who (in contrast to Moritz Schiff and Giuseppe Colasanti) found that fatty degeneration and either atrophy or loss of the specific properties of the olfactory epithelium occurred after cutting through the olfactory nerve of a frog.

We are unable to offer any decisive observations about the degree of qualitative changes due to increased use, especially about whether the increase in specific performance is steadily progressive or, more likely, reaches a maximum after a short exercise, whereby its significance for the gradual differentiation of the organs is only very slight. However, something theoretical, for or against, will emerge from the following considerations.

Even if the effect of qualitative functional adaptation is limited, both it and quantitative functional adaptation are of the greatest importance for animals which, without them, would forever remain at the level of the innate and bequeathed. Then, we must remain like newborn children in all our abilities and capabilities. Friedrich Schiller's justified words in *Wallenstein,* "It is the spirit that creates the body," would have no meaning.

2. Functional Self-Organization of Purposive Structure

After the transformative effects of increased or decreased use have been briefly discussed in an analytical manner, we must cite a group of forms which, in their causes, closely follow these changes and also have much in common with the phenomena mentioned in terms of their heritability.

The phenomena, discussed earlier, of the effects of the frequency and intensity of use are of greatest physiological importance as determining all quantitative relations in the body but have been undeservedly neglected by physiologists (with a few honorable exceptions): probably because these effects are not observable in the brevity of physiological experiments, only become sufficiently apparent over the course of years, and can in part only be ascertained by statistical means. In a similar manner, the phenomena to be discussed now have been completely disregarded by proponents of the doctrine of descent, although they are of fundamentally decisive importance for that doctrine.

These are phenomena that can be summarized, with the previous phenomena, under the common name of functional adaptation. What has been discussed previously are effects of the quantity of function on the external form and quality of organs. The phenomena that will now be discussed show us the effects of function on the inner form or "structure" of organs. We can summarize them both as *"principles of functional self-organization of the purposive,"* the former pertaining to external shape, the latter as influencing the internal shape of organs. Both sets of phenomena are based on the principle that function directly produces what is purposive. It goes without saying that they are closely interrelated.

We owe the first observations pertaining to this internal self-organization to Hermann Meyer, who recognized that the spongy substance of bones has a very specific architecture, which at every point exactly represents the lines of the strongest pressure or tension to which the organ is exposed. Because the trabeculae run everywhere in the directions of greatest pressure and tension, the greatest possible strength is achieved with the least amount of material, exactly as modern constructive technology seeks to achieve. Our relevant knowledge was then expanded by Julius Wolff, Hermann Wolfermann, Karl Bardeleben, Friedrich Merkel, Christoph Aeby, and Paul Langerhans, and thus extended to almost all bones of the human body and of some mammals. [William Marshall found a similar static structure in the beak of birds and Hans Strasser in the bones of the trunk and extremities of birds.]

Julius Wolff was the first to discover (confirmed by Ernst Küster, Erich Martini, and Ludwig Rabe) that such structural relations also develop under completely novel, abnormal conditions that correspond to new static conditions as found, for example, with crookedly healed bone fractures and rachitically bent tubular bones. This shows that these formations cannot be fixed and bequeathed but can always generate themselves according to the respective circumstances. Because the static bone structure only becomes recognizable after the first few years of life, nothing can be said about its possible heritable transferability without special investigations aimed at this question.

Furthermore, a personal communication from Karl Bardeleben, made two years ago, belongs here, which I cite with his permission. He spoke of the assumption and probability that the fibers of the fascia (the sheaths that surround the muscles) assume the directions of strongest traction, as occurs for the trabeculae of bones. Because he has not yet undertaken the intended special investigation, I have, without wanting to anticipate his special communication of this work in any way, convinced myself of the correctness through my own observation, at least as far as to be able to confirm and evaluate it here. I must add that a year ago Professor Hermann Meyer expressed to me the same thought and an intention to extend investigations from this point of view to all formations of connective tissue. Without wanting to anticipate the publications of this author either, I simply state that I consider this intention to be highly justified because why should not, for example, the direction of the tendon fibers or the fibers of the *ligamentum interosseum antebrachii,* which always lie in the direction of tension, be understood in the same way?

It appears to me that one of the most instructive examples of these relations is presented by the familiar fibers of the eardrum. The two main fiber systems, the radial and the circular, simply substantiate those directions that must withstand the greatest stretch during vibrations of the eardrum. However, a third system is clearly developed, which transmits vibrations of the eardrum to the inserted long extension of the hammer. These fibers are again oriented in the direction of strongest tension—that is, at right angles to the extension of the hammer. [The wording of these last sentences was not entirely correct. They should have read: In the area of the "course" of the long extension of the hammer inserted into the eardrum, the radial fibers that attach to the free "end" of the extension are deflected from the radial direction in the most favorable direction for the transmission of vibrations to the extension, again in the direction of the greatest tension.

The semilunar valves of the heart also have typical, but simpler, structure adapted to their function; less so the valves of the veins.]

With the exception of the two passively functioning organ systems— bones and connective tissues—we recognize comparable structural relations, which can be deduced from the same causes, in the third mechanically but actively functioning system of muscles. The relations appear simple for the skeletal muscles, at first glance almost self-evidently simple, but they are not so simple everywhere; and I will save special information on this matter for the end of a special investigation directed to it.

On the other hand, it has long been known that in cylindrical hollow organs (such as the intestine, ureter, and blood vessels) smooth muscle fibers run in only two directions—the longitudinal and transverse directions of higher-achieving function—and we therefore have a right to include them here. The same is true of vesicular organs in which the fibers run just in the equatorial and meridional directions, again in the directions of strongest performance.

One of the organs with striated muscles also belongs here. When viewed in this way and once the principle has been established, the fiber directions of the heart promise to instruct us about the nature of the heart's function and the directions of its greatest achievement in action. [Also to be mentioned are the muscles of avian stomachs and mammalian tongues that are equipped with three orthogonal systems of muscular fibers.]

All these formations of bone, muscle, and connective tissue could never have been selected with such regularity and perfection from individual variations of form, according to Darwin, because thousands of fibers or trabeculae must already have been appropriately ordered in order to bring about only the slightest advantage noticeable in the budget and cultivated by selection through material savings, and because in starvation it is precisely these parts (excepting the heart) that would suffer last because of their low material exchange, much later than other, more living organs with higher material exchange.

These formations therefore cannot be derived through selection of individual variations of form, as is the basis of Darwin's doctrine, but can only be derived from the "creative reactive qualities" of the tissue in question, which directly shape what is purposive down to the last detail. These qualities are set out in this treatise. In the following chapters, we intend to expound the necessity of their origin and factuality of their existence.

The related formations of the connective tissue organs and of the sheaths formed from smooth muscle fibers are already congenital, and could therefore be regarded as evidence of the heritability of functional adaptations.

But we shall see in the special investigation of heritability that this conclusion, in spite of this congenital occurrence, cannot be drawn.

In addition to these static adaptations of the inner structure of supporting organs and the dynamic adaptations of muscle fibers to directions of highest performance, there is another group of forms to be mentioned that achieve the highest possible performance with the minimum of material. These are the shapes of the lumens of blood vessels, which have the same character with regard to performance but differ from those others only in that the forces to which the forms adapt are not static (as in supporting organs), nor simply dynamic (as in muscles), but much more hydraulically complex, specifically hemodynamic.

The nature of these relations is generally as follows:[8] the lumens of blood vessels at the origin of each branch do not have a cylindrical shape as in the course of the vessel but possess the peculiar conical shape that a jet automatically assumes as a result of the hydraulic forces acting within it as it emerges unchecked from the lateral round opening of a cylinder through which it flows. The shape of blood vessels changes in exactly the same way, under the same conditions, as the shape of such a freely emerging jet; this modification therefore takes place with changes in the size of the angle of inclination of the branch to the pipe through which it flows, with the strength of the branch in relation to the strength of the trunk, etc.

At the same time, this implies that branches emerge from the trunk in the direction determined by the velocity of flow and the magnitude of the lateral pressure, and from this orientation they only gradually turn toward their area of circulatory distribution.

If, furthermore, an arterial trunk gives off branches that are more than two-fifths of the trunk's diameter, the trunk itself experiences a deflection to the opposite side, and this deflection again increases in accordance with the hydraulic conditions, with the size of the branching angle, and with the thickness of the branch in relation to the strength of the trunk.

All these devices have the consequence that blood is distributed through the body via innumerable ramifications with the least amount of friction

---

8 This was the topic of Roux's 1878 doctoral thesis; see Kurz, Sandau, and Christ (1997). Roux's thesis also was published in the same volume of collected works in which he published the second edition of *Struggle of the Parts*. —Trans.

so that the operation of the circulation is made possible with a minimum of living force and a minimum of wall material. Other properties of the shape of the blood vessel lumen have long been known and act in the same way, such as the perfect smoothness of the inner wall, the cylindrical nature of the lumen in the course of the vessels, and, above all, the shaping of the main paths of the reticulated system.

All these properties are present at birth and, apart from the directional relations, develop by themselves even under abnormal conditions. This indicates a truly wonderful property of the blood vessel wall. In order to enable the bloodstream to create the described conditions everywhere by itself through the forces contained in it, the blood vessel wall must have the property of resisting strong blood pressure while at the same time yielding completely to the impact of the finest liquid bursts. If the walls of blood vessels possess these properties, all the forms listed above (as well as forms not mentioned here for the sake of brevity but described in my earlier works[9]) result completely automatically! Moreover, the wonderfulness of the properties of blood vessel walls are considerably increased by their ability to withstand even the strongest surge of liquid in places where this is necessary for the organism.

These three properties (resistance to strong pressure, yielding to fine bursts, withstanding strong bursts) would contradict each other for a dead substance. And yet it seems more natural to ascribe these properties to the living wall than to allow each and every one of the millions of ramifications of the circulation to arise through individual formal laws, which moreover would provide no explanation for the formation of the same arrangements in abnormal novel conditions, for example, after the ligation of arteries. From this adaptation to novel conditions it follows again, as in the two previously discussed formations (of supporting tissues and muscle fibers), that the respective forms could not have originated and been cultivated by individual variation and selection. This is quite apart from the fact that this cultivation would have been impossible because the coincidental occurrence of some of these variations in the struggle for existence would have

9 Wilhelm Roux, "On the Ramifications of the Blood-Vessels in Humans," inaugural doctoral dissertation, Jena 1878; "On the Significance of the Deflection of the Arterial Trunk in the Branching Process," *Jenaische Zeitschrift für Naturwissenschaft* 13 (1879): 321–337.

been of absolutely no use, and besides, an accidental occurrence of such forms with their delicacy (compared to which the architecture of spongious bone is building with rough timber) belongs entirely to the realm of improbability, for the characters of the branch-origin-skittle[10] are so fine that they are often completely lost when drawn due to a deviation of just the width of the line.

So these forms show the existence of qualities in the organism that, upon the action of functional stimuli, are able to directly produce (directly form) what is purposive in the highest conceivable perfection. From where do these wonderful properties come? Emil Du Bois-Reymond asked himself this question years ago when he wrote, "But the ability of organisms to perfect themselves by practice does not seem to me to have received sufficient attention with regard to natural selection."

Does not this ability reintroduce teleology and thus the dualism happily defeated by Darwin? The next chapters will attempt to answer this question.

B. Heritability of the Effects of Functional Adaptation[11]

1. *The Facts*

The extent or breadth of an individual's functional adaptation is known to be limited. As concerns the acquisition of bodily skill or mental perfection, each individual can raise himself only to a certain level by his own diligence whether it concerns the acquisition of physical skills or mental perfection.[12] These changes which are very beneficial for the individual, however, would have been entirely useless for the development and perfection of the whole animal kingdom if they were not bequeathable (transferable to the offspring).

---

10 *Astursprungskegel* is Roux's term for the smooth shape where vessels branch. —Trans.

11 Heritability is qualified by [*eventuelle*] (possible) in the table of contents of the second edition but not in the subheading as it appears in the main text. —Trans.

12 *Geistige Vervollkommnung* could also be translated as "spiritual perfections." —Trans.

And if this inheritance was not able to raise the offspring to a higher level from the outset, each offspring taking a farther step could work its way up to even higher perfection with the help of individual adaptation.

The degree of progress possible in this way would depend on the speed of bequeathal of these acquired, purposive characters. If every acquired character were transmitted to offspring, then progress could be exceedingly fast. However, the slowness of progress indicates that only a small proportion of acquired characters are transmitted. Indeed, it seems as if persistent action of functional adaptation in the same direction for many generations is necessary for characters to be fixed in this way and to be transmitted to offspring by bequeathal.

When determining whether an acquired property has been bequeathed, one is always faced by a choice between two options that can almost never be decided with certainty. This is the question of deciding whether the inherited favorable quality has in fact been completely newly acquired by the father and then bequeathed by him, or whether it had not already occurred in him through embryonic [better, according to August Weismann, blastogenic, germ-plasmatic] variational potential and only "developed" by him in later life. And of these two possibilities the second, which is unfavorable for development,[13] seems almost always to be the more probable.

That novel embryonic variations are very often, and to a high degree, bequeathed can no longer be doubted by anyone, even if there are cases in which embryonic variations are not bequeathed such as, for example, asymmetric gigantism which is always congenital and many tumors with congenital germs.

One can rely upon someone to raise the following objection to explain the high level of industrial skills, which are observed to have increased from generation to generation in regions where almost the whole population has cultivated the same industry for many generations. Someone can always make the not unwarranted objection that only those siblings were trained to continue their father's trade who, from their youth, revealed a particular

---

13 *Entwickelung* is probably intended here in the sense of phylogenetic, rather than ontogenetic, development. —Trans.

aptitude for the trade, who had been born with skills acquired through accidental embryonic variation. Continued selection in this one direction through these many generations could then account for the intensification of performance, apart from any expansion in the breadth of individual adaptation caused by early youthful employment.

Most authors have contented themselves with expressing subjective opinions about the heritability of functional adaptations; only a few have supplied factual material.

Darwin first indicates the important fact of the bequeathal of instincts. Although many instincts can be thought of as originating from embryonic variations, such as the olfactory instincts, there are also some that could only be acquired through personal observation and experience—that is, through functional adaptation. Thus, Darwin directs us to the acquisition of the fear of humans by animals. When humans come to hitherto uninhabited islands for the first time, the animals often have no fear of them; but after several generations, fear of humans has become a congenital instinct. Furthermore, Sigmund Exner continues, "Not only is memory, as the ability to hold memory images for a longer or shorter period of time, bequeathed, but also the content of memory, the memory images themselves. It happens that young hunting dogs that have never hunted nor had any other opportunity to get to know the firing of a shotgun and its effect when they saw the first shot in the field, rushed with full pleasure, like an old hunting dog, to retrieve the prey even if they saw none falling. This proves that since the invention of gunpowder, the memory-image of a shot and its consequences has become hereditary in the dog's brain, that is, has been inherited as a so-called instinct."

Further examples of the acquisition and bequeathal of instincts can be found in the works of Ewald Hering, Victor Hensen, Ludwig Büchner, Karl Schneider, Arthur Erwin Brown, and others.

The fact, on the other hand, that the inheritance of concrete content of the mind is so limited in humans is striking, but must be regarded as a very favorable quality especially acquired and cultivated in the struggle for existence, since, as is well known, our universality is the cause of our main advantage over animals; for if we inherited the knowledge of our ancestors in the same way as animals, the freedom of individual training would thereby be restricted as it is in animals.

It seems conceivable that this property took its starting point from a reduced congenital disposition to bequeathal of mental contents and was then further developed through the great changes in the way people live because the acquisition of instincts probably requires repetitive impressions over many generations, combined with a certain simplicity and limitation of the whole mental contents.

In his report of the hereditary transmission of crooked baker's legs, Ludwig Overzier provides an example of the bequeathal of an acquired character, which because of its nonpurposiveness could not have been cultivated by selection.

I have striven to increase the number of assured examples of the unambiguous bequeathal of functional adaptation. These are restricted to such qualities, which either did not arise from accidental embryonic variation or cannot be assumed to have been cultivated by selection.

In my opinion, the congenital disposition toward the mother tongue cannot have arisen through embryonic variation. The coordination, arrangement, and connections of the ganglion cells, which innervate the muscles of speech, are congenital to the extent that we learn to speak our mother tongue most easily. Thus, Europeans brought up as children with Namibians do not learn their language, or only do so with great difficulty, as compared with the ease with which they learn their own language.

The coordinated eye movements, which always position both eyes for every direction of gaze so that the images of each object fall on identical points on both retinas and are therefore easily seen, are inherited because, according to the investigations of Eduard Raehlmann and Ludwig Witkowsky, they are already the predominant ones in the first ten days of life. This does not contradict William Preyer's observations that they are not immediately congenital but only acquired within this time, but it proves that at least their disposition must be congenital. Such infinitely complex connections of the muscular movements cannot, in my opinion, have arisen from accidental embryonic variations.

Better still, the following consideration seems to me to be even more conclusive than these two examples.

As mentioned, the question at hand is always a matter of differentiating what has arisen through accidental embryonic variations and selection according to Darwin's principle of selection, from what has developed through functional self-organization and is then bequeathed. The effects of the first principle appear unlimited; we can hardly recognize any changes, however great, for which it could be asserted with absolute certainty that they could not have, in principle, arisen through embryonic [blastogenic] variations and selection, sufficiently repeated, provided that selection worked finely enough over sufficient time. Nevertheless there is an incident in the development of the animal kingdom of which the opposite appears reasonable.

For there is a period in the developmental history of the animal kingdom for which we can affirm with certainty that perfection was not gradual or "successive" in the individual parts but must have been "simultaneous" in almost all organs of the body. For if there had only been favorable variations in individual parts, life could not have progressed beyond this point. It is a period in which the development of a thousand, even a million, useful individual properties must have taken place all at once and which cannot have been achieved by selection from free variations that did not already tend toward the purposive because selection can only generate characters gradually, a few at a time. Now what was this momentous occasion for which we can assert this necessity, for which the transition cannot have been gradual or successive in the various organs? It was the moment of transition from water to land or, more correctly, to life in air. We are accustomed to regard this transition every year in young amphibians as something quite natural; but here the changes of the animal in all its parts, like all other embryonic transformations, take place according to certain inherited laws of development, and the transformation of a tadpole into a frog, in this manner, is nothing special. But how were these laws of development acquired? How did these properties first arise when a thousand or a million changes were needed all at once? Perhaps they were not so many and perhaps a gradual reorganization was possible for this adaptation. No problem! Gradually! Adaptation was gradual. At first, animals stayed on land only briefly and soon returned to the water. But what is necessary for an aquatic animal to live on land even for a short time?

Let us consider this process in the attempt of vertebrates to venture onto land, and let us see how this venture would have affected the body and what was necessary for them to succeed. Let us grant these animals, in addition to gills, a vascularized swim bladder that has been converted into a lung, acquired by earlier gasping for air with the aid of selection.

As soon as the animal comes out of water onto land, it would feel the most terrible discomfort, for suddenly its body and limbs would feel much heavier than before because they were no longer supported by the weight of displaced water. How uncomfortable do we feel when we climb ashore after prolonged swimming and suddenly must again support our bodies' weight? The minor discomfort we experience in this transition, accustomed as we are to carrying our members all of our life, is not at all comparable to the feeling that an animal must experience which has never before had to support its body parts by itself.

Furthermore, the animals must at once move quite differently and coordinate their muscles differently. They can no longer move as they had been accustomed to move in water, where they were almost weightless, but must use almost all their muscles very energetically in a manner prescribed by the statics. Furthermore, the bones, which hitherto had to offer resistance almost solely to muscular action, must now suddenly bear weight according to the static conditions—indeed, so strongly that carrying the body in water can hardly be considered preliminary exercise for walking on ground. The same is true of the joints, the cartilage, and ligaments; they are all suddenly subject to greater demands, the ligaments in new principal directions.

The distribution of blood in the body is immediately quite different: the blood, which hitherto was shielded from the effects of gravity, now sinks from the brain and spinal cord to parts of the body nearer the earth. The mechanisms that regulate the flow of blood to the various organs must immediately distribute the blood according to completely new rules if a total disturbance of the functions of all organs is not to occur, especially a paralyzing anemia of the central nervous system.

The body will be starved of oxygen because the lungs must now meet all needs for a longer period of time. Abnormal sensations will be felt as the skin, gills, and appendages dehydrate. The hitherto safe intercourse with the outside world is abolished, for the sense organs cease to function as they receive completely new impressions that have not become comprehensible through experience. The organs of hearing, accustomed to stronger conduction through water and the transmission of impressions through the whole skull, will hardly be spoken to. The eye will have lost its ability to form images. Whether the loss of heat by evaporation of water will be disadvantageous in these cold-blooded animals remains to be seen.

These evils will in part grow with the duration of sojourn on land, and the visit will therefore initially only have been very brief, and these evils will also occur in the protruding parts if there was only partial emergence from water. But what is most important, these evils always occur all at once, and if the animal nevertheless ventures onto land, their correction must also occur all at once.

What does such a simultaneous correction in all organs of the body, except those of nutrition and reproduction, signify? It signifies the presence in almost all parts of the body of the most perfect mechanisms of functional adaptation that are able to bring about the necessary, purposive changes directly in the transition of the organism into new relationships. They are an indispensable precondition for even a temporary exchange of aquatic for subaerial life, a necessary requirement that becomes all the more imperative the longer the stay on land continues.

We know such self-regulative mechanisms from the higher animals, and from this we conclude that lower animals probably also possess them. We know our ability to acquire completely foreign modes of movement and through practice to make them easily practicable, to transform all the central motor organs in the brain and spinal cord accordingly. We know that the function of the bones and ligaments becomes stronger in the places concerned. When the body is in a normal condition, we can convince ourselves every day of the precision of the regulation of blood distribution when we stand up from bed in the morning without noticing even a momentary anemia in the brain. In pathological disturbances, respiration regulates itself to a very considerable extent, and Karl von Schreiber has observed that when the proteus lives in shallow water its lungs become larger and more vascular, and its gills correspondingly smaller.

Regarding the degree of direct adaptability of the sense organs, we cannot allow ourselves to draw any conclusions from the higher animals as to what was required here. However, the diminution of the function of these organs was at first less disadvantageous because there were at first no enemies on shore.

Thus, it was necessary that purposive changes took place at the same time in almost all organs. The question is whether functional adaptation can achieve this or whether this contradicts its essence. We shall further explain in detail below that this is precisely its essence, just as it is able to act appropriately and remodel in millions of individual places in the same organ or organ system.

It must be emphasized that in response to a change of living conditions, functional adaptation can produce equally useful changes in "all" affected organs of the body "at the same time." This simultaneity of effects in millions of parts is its characteristic feature relative to selection which can develop only a few useful properties at a time.

Now we can further investigate the hereditary nature of the effects of functional adaptation.

Let us first assume that the effects of functional adaptation are not hereditary. In this case, every generation that tries to look for food or protection from enemies on the bank outside the water will have to start from the same stage and therefore will never be able to exceed a certain level of perfection in adapting to life on land, because exercise has its definite limits for the individual. However, over the course of generations, fortuitous, congenital variations will gradually occur and perhaps give their bearers an advantage. It must be taken into account, however, that this can only be very slight because the more favorable properties exist only in some parts, but simultaneous corresponding changes are necessary in all parts; it is possible that for this reason an advantage may not come into effect at all. But let us assume that it comes into play; so this animal would proceed a little farther in adaptation, and, as this repeats itself, perfect adaptation could gradually take place through variation and selection, and functional adaptation would have merely facilitated the transition.

But if we now ask what are the accidental, congenital, and therefore hereditary properties that would have been accumulated through natural selection, we find that they are almost exactly the same properties that functional adaptation already has formed during all stages of the transition. However, because of their lack of heritability these adaptations would not have been transferable to the offspring. So all these millions of changes, which the individual acquires at once to some degree through functional adaptation, would be lost and must be acquired again and fixed anew little by little by means of the infinitely wide detour of the selection of arbitrary variation. And this must not have taken place once for each part but must have been repeated for each part step by step up to the degree of perfect adaptation. From the fact that the finer structure of almost all organs of the body must more or less be changed, it follows that we did not exaggerate when we spoke of millions of individual properties; we would have spoken more correctly of billions.

If the effects of functional adaptation were not hereditary, that which functional adaptation would have created purposive in a thousand parts of the organism all at once would have had to have been acquired and acquired over and over again in hereditary form through thousands of generations of accidental variation and selection, not just here but everywhere in the further development of the organs. If, on the other hand, the formations of functional adaptation are transferred to the offspring as soon as they have been acquired and preserved for several generations, then in this manner a great number of the purposivenesses of animals find their [simplest] explanation, provided that functional adaptation itself is explained. It is understandable that this inheritance is very small in humans because almost every generation has a different occupation and way of life. Our uncommon versatility of individual activity impedes the development of fixed mechanisms and also makes their bequeathal difficult. Therefore, we find only those functional adaptations bequeathed that are constant despite general change: coordination of the mother tongue, coordinated eye movements, and the most general concepts of space, time, and causality.

The coordination of speech and eye muscles, if they are to be of any use, must always take place in so many thousands of ganglionic connections that these cannot arise through the accumulation of accidental embryonic variation by selection; and if the disposition for this formation must therefore be assumed to be congenital, it can [in a simple way] result only from the bequeathal of acquired functional adaptation.

2. *Theoretical Considerations about Bequeathal and Development*

After this empirical treatment of bequeathal, we want to add a few theoretical principles.

First of all, an error must be pointed out that is often made in judging the heritability of developed forms. Many consider those, and only those, forms as inherited that are regularly congenital. However, this view must be described as incorrect in both directions. Not all regular congenital forms can be regarded as directly bequeathed, nor can all forms that appear after birth be interpreted as acquired and not bequeathed.

If the former were correct, if all congenital formations were bequeathed, then in the above example of the congenital functional structure of the connective tissue and arteries we would have one of the best proofs of the heritability of structures produced by functional adaptation, and one would probably at first glance be inclined to use it in this way. Unfortunately, this would be incorrect; for this conclusion was based on an incorrect, superficial conception of what can be bequeathed.

The moment of birth cannot be viewed as a dividing line between what is inherited and what is acquired. In truth, fundamentally new conditions come into being at birth only for the respiratory and digestive organs; all other organs function to some degree in the womb and thus have already been affected by functional stimuli.

The movements of the embryo in the womb are known to all. William Preyer's most recent investigations, his embryoscopy of hens' eggs, reveal that such embryonic movements occur in the very earliest stages in the most persistent manner. He observed that the chicken embryo moves its trunk and extremities in a lively rhythm from the third day of incubation.

Therefore, from their earliest formation muscles with their associated bones, aponeuroses, and fascia, as well as skeletal parts with their joint ends, capsules, and ligaments, are under the formative influence of their function. We are therefore not justified to regard formations present at birth as being directly inherited. We are unable to judge how much is "directly bequeathed" and how much "acquired" through functional adaptation because we do not know the functional range and speed of embryonic adaptation. For these reasons, we are not yet able to distinguish "directly bequeathed" from "secondary" formations.

From what follows, it will be seen that relatively few characters need to be directly bequeathed,[14] preferably only those that originally arose through embryonic variation and then produced the specific individual forms with the directing help of functional adaptation.

---

14 The "directly bequeathed" formations represent the part of evolution, the secondary formations the part of epigenesis in individual development.

The internal structure and external form of the blood vessels present at birth cannot be regarded as bequeathed (even less than for the muscles, skeletal parts, ligaments, and fascia) for blood vessels function continuously in the embryo from their first rudiments; and the structure of their walls and the shape of their lumens at birth can thus arise through functional adaptation in the same way as the pedestrian unequal development of organs in adults.

The sense organs are more or less struck by stimuli that can take part in the formation of their percipient parts, although this effect is usually only slight.

The same applies to the development of birds as of mammals. Here, too, there is a fixed point in time when an obvious change in the conditions of life occurs, but the moment of crawling out of the egg is not considered sufficiently impressive to be the boundary between what is inherited and what is acquired. If this view of a borderline is entirely unjustified in light of the above, then the question must be asked: where is the corresponding borderline in amphibians and fish, which from the outset live almost as if in the open air and are relatively little protected from the stimuli of the outside world by their egg shell? Who would dare to fix a moment here when the bequeathed formations cease and acquisition of properties through functional adaptation of the embryo begins! At what relatively early stages of development are these animals already dependent on self-nourishment! Could the boundary between the bequeathed and acquired be placed at the time when an animal begins isomorphic growth of all its parts? Then, by analogy, one must describe human life almost until adulthood as embryonic or bequeathed because, as is well known, true isomorphic growth does not exist; rather, the various organs grow unevenly at each developmental period. As for external proportions, artistic anatomy teaches us that every age is characterized by distinctive proportions of parts of the body; and weights of internal organs at different ages show the same thing. Detailed morphological examinations confirm this in all organs.

Now where does bequeathed growth end? All these unequal changes in the organs in the various periods of development up to adulthood are evidently bequeathed, insofar as they keep within the fixed norm.

Nothing can be decided here as long as we remain unclear about what is meant by bequeathed formations. At least we can say that it is arbitrary to exorcize the origin of bequeathed formations from the embryonic period and to designate all postembryonic formations as acquired. Embryonic, from ἔμβρυος, that which is enclosed (in another body), designates only one single circumstance of an event, that of a certain isolation, and certainly everything that takes place during this time can be called embryonic. But first, as mentioned, this isolation from the outside world (from external influences) is very imperfect, and second, as we have shown, this period by no means coincides with formation of the bequeathed. If, however, we want to use the term "embryonic," as is habitual, as synonymous with "bequeathed" because the formations that take place in the embryonic period are for the most part bequeathed, then according to the principle *a potiori fit denominatio* we must not shy away from identifying boyhood as largely embryonic.

"Bequeathed" is understood in the ordinary sense to mean properties that an individual's ancestors already possessed and passed on without impediment to their descendants, just as a father's fortune acquired through work is bequeathed to his children without the child having to do any of the work that the ancestors performed in order to gain it. This definition seems to me to capture the essence and to be well-suited to be transferred to biological events. New properties come from parents through the parents' activity, by their adaptation to functional and other [atypical, external] stimuli. By the biological growth of assets that they have inherited, parents bequeath to their offspring an endowment that has grown. Only those properties are "inherited" that are transferred to the children "as they are" without functional fashioning [willed or induced by external circumstances] by the children, and thus are fashioned purely from internal typical causes.

But because, as mentioned, many muscles function in the embryo, the parts dependent on them (tendons, bones, joint capsules, ligaments, and fasciae) are subjected to unconscious activity and are therefore compelled to develop dependent properties. If at the same time the muscular arrangement is perturbed by pathological influence, no one would expect to find the dependent parts to have developed in a normal manner, and this is what actually happens in malformations according to Antonio Alessandrini and Ernst Heinrich Weber. These authors found that if nerves of a corresponding nerve area were absent in the spinal cord, then the muscles were also absent. Moreover, the associated bones and joints were abnormally formed, the latter in part being stiff. Weber found tendons and tendon membranes to be present but does not report whether they were completely normal, and it seems highly unlikely. If, however, the spinal cord was originally formed but destroyed by disease in later embryonic life (*spina bifida*), then the parts of the musculoskeletal system were found to be perfectly normal, and it must then be left to further investigations to determine how far active embryonic function of the muscles or only the tone of the same is necessary for the normal development of their supporting apparatus.

Likewise, in the absence of the tendon of the long head of the biceps, Georg Joessel found that the *sulcus intertubercularis,* through which it runs, was only weakly developed and the synovial capsule was not everted into this sulcus, a statement I can confirm from multiple observations of my own. The dependence of the development of the "passively" functioning parts on the embryonic functioning of the "active" parts, as demonstrated in this way, is also in agreement with the investigations of Hjalmar Heiberg who found that the joint capsules of the newborn are even stronger and tighter on the flexor and extensor sides than in adults whereas the accessory ligaments are weaker or still completely absent (and other features not to be listed here). August Förster describes an orbit (eye socket) without an eye; but it was abnormal, narrower than the orbit with an eye.

From these examples it appears that the structures of the supporting substances are "independently laid out" but only attain their normal development with the cooperation of the parts they support under the influence of their function.

Examples of other dependencies (perhaps also more functional) are as follows. If some of the fibers of a muscle in the embryo become aberrant, the associated nerves, blood vessels, and tendon fibers also vary in the corresponding manner. According to Justus Carrière, if the ganglion of a snail's tentacle is cut away at the same time as the tentacle and eye, then no new eye grows again; but if the ganglion is retained, the eye regenerates in the most perfect way.

It does not follow from the foregoing that the passively functioning parts (supporting structures) arise in absolute dependence on the active parts.

Rather, it seems possible that the relationship can occasionally be reversed. For example, a change in the bones could have arisen primarily in the embryo by embryonic [blastogenic] variation, have been cultivated by selection, and only secondarily resulted in changed muscular forms. In other words, the change in the supporting structure could have led to a change in the use of the extremity and thus to a corresponding deformation of the muscles through functional adaptation.

As mentioned, the same is also true of the blood vessels. After their initially self-sufficient installation, their further development depends on their continuous function; if the organ to which they pertain—for example, a kidney—is lost, the blood vessels do not continue to develop normally in the same way as if the kidney were present. They are dependent formations that receive their normal size and shape through functional adaptation in the embryo, not independent formations that are developed and cultivated by fixed bequeathal. The fact that blood vessels occasionally grow independently to form tumors such as telangiomas (red birthmarks) and cavernous angiomas (venous vascular tumors) cannot be regarded as arguing against this. For we know just as little with certainty (for these parts as for all the others) about what stimulates them to particular growth and what prevents this. For it is precisely the parts that depend on blood vessels for their shaping (the connective or supporting substances, bones, cartilage, and connective tissue) that most frequently exhibit "tumorous transformations," while the active ganglion cells, nerves, and muscles do this only rarely, probably because they come too quickly and too completely under the control of functional stimuli, as will be discussed later, so that without the functional stimuli they cannot live and grow, and therefore cannot form tumors.

The eyes of an adult snail will grow again if we cut them off, even if the snail is kept in the dark. The regeneration of an eye is therefore an embryonic event that takes place in the adult, for the act of cutting cannot be regarded as the intrinsic cause of the formation of an eye but merely an occasional cause or accident. The eye takes shape without any external cause, owing to the internal properties of its parts. In addition to its qualification as the bearer of the eye, the tentacle has also retained the embryonic properties for the formation of an eye. The cells, or just particular cells, of these animals perhaps contain, whether in their nucleus or in its vicinity, *"unmodified remnants of embryonic substance"* [reserve idio-plasm] which exhibit their formative properties in the event of defects.

After what has been said, we shall continue to designate as bequeathed {or embryonic} only that which formally or chemically differentiates itself or even becomes larger from purely internal causes without any differen-tiating action outside the part itself.[15] And that which grows and differen-tiates itself by its own power already has the power to attract and assimi-late more food than it consumes. Thus, growing by one's own power is called "embryonic," in contrast to the normal enlargement of organs in adults, which, as we intend to explain in later chapters, can only be stimu-lated to further growth under the action of functional or other stimuli. In certain circumstances, the connective substances could be considered embryonic because they are able to enlarge under their own power if sup-plied with more blood by the action of stimuli.

The consequences of this view will be explained and justified in detail below. Here we want to go backward to discuss the development of bequeathal and then the essence of bequeathal itself, as much, or more correctly, as little as it seems to us to be conducive with the knowledge of our time.

The preconditions of development, or essential properties of the or-ganic, are material exchange and shaping by chemical processes.[16] Both

---

15 The qualification "or embryonic" was present in the first edition but deleted in the second. A footnote in the second edition reads, in part, "I later called this manner of forming parts self-differentiation. . . . One could perhaps distinguish forms that arise through typical 'dependent' differentiation (i.e., by epigenesis), as indirectly bequeathed [*vererbte*], from forms that are directly bequeathed [*vererbten*] by self-differentiation (evolution), just to give names to the different modes of inherited [*ererbten*] forms." —Trans.

16 [The statement made here of a basic principle of organic formation "directly" by "chemical" processes is untenable and was soon abandoned by me. Incidentally, it is repeat-edly deviated from in the following pages. . . . I soon recognized that the bequeathed sub-stance [*Vererbungssubstanz*] must have a specific structure that establishes a continuity of or-ganic formations from ascendants to descendants, in that this substance is transmitted uninterruptedly from individual to individual. I had already developed these ideas myself when I heard August Weismann present, and carefully justify, the idea of the continuity of the germ-plasm at the meeting of naturalists in Strassburg in 1885; the same idea had already been expressed briefly in 1880 by Moritz Nussbaum and even in 1876 by Otto Bütschli in a less well-known place. . . .]

are incomprehensible to us, but the latter phenomenon is most perfect. [Both are based on assimilation as defined and discussed previously and in the following.]

The essence of material exchange consists in the fact that in the course of organismic processes the constituent parts are changed in their chemical arrangement so that they are unsuitable for the further progress of the process and must be removed, while at the same time the capacity is developed to attract, from the neighborhood's zone of molecular attraction, differently grouped parts and to regroup them as things similar to themselves. The latter process is called assimilation, the former dissimilation (Ewald Hering). There is nothing wonderful about dissimilation. It is continually observed in inorganic processes. Assimilation, on the other hand, is less understandable. It can be compared to the training of the recruits in a regiment. New men are always being trained, "assimilated," by the branch of service, and this is done in a different way in the regiments of each branch of the armed forces, while the old or dead leave the unit again and again. Viewed as a whole, the material changes constantly; what remains and is instantiated is merely the abstract regiment, represented by its statutes. Just as in a school, the same thing is taught to the students, according to the regulations, by quite different directors and teachers in the course of time; and the number of teachers, the assimilators, is always relatively small compared with the number of students, the assimilands.

But the development of an embryo is different. Here a new factor is added to assimilation. Very few instructors, the constituents of the fertilized egg, take in many new students from without and assimilate them. But the new, wonderful thing is that the instructors continue to change and so do the students. The statutes are therefore not fixed, but are different for teachers and students in each subsequent period. We do not know whether the statutes are followed by the teachers first, who then assimilate the students, or whether the statutes educate teachers and students at the same time. It is like a journey through elementary school, folk school, grammar school, and university one after the other, but with the peculiarity that the students immediately become teachers and enter the next higher school as students, assimilate there, and become teachers before going to the next higher school as students. Here the teacher is probably only a little ahead of the student. On the whole the same thing happens in our schools; but the riddle in the development of the egg is through what and how from the original statutes of the elementary school, the statutes of the folk school, grammar school, and university develop one after the other. Quite apart from the question of how the statutes affect the teachers, we assume that the material of the egg in its specific constitution already represents the trained teachers of the elementary school.

But it is clear and self-evident that these cannot be simple elementary teachers if they possess the ability to develop into high-school teachers on their own, to prepare their students to become high-school teachers in the near future, while at the same time training themselves to become university teachers. Elementary instruction would probably be taught differently from the beginning; it would already have something of the purified spirit of the grammar school in itself—for example, as if a grammar school teacher gave elementary instruction—and the ability for further development must already be present *potentia*. It is this necessary difference from the outset that Wilhelm His and others rightly oppose to the basic biogenetic law of Fritz Müller and Ernst Haeckel. As His points out, it is impossible that an egg that contains the chemical constituents for the later development of a human being can at any stage really resemble an egg that is capable of developing into a bird or amphibian. The necessary structural and chemical difference cannot be completely ineffective for a time. It will not only lead to abbreviations of bequeathal but to real deviations from ancestral states. Because the morphological process can only be derived from the chemical process, *these differences somehow must assert themselves morphologically in an earlier period,* whether or not we able to recognize them, quite apart from special adaptations to the sexual organs of the mother which are generally recognized.[17]

The version of the basic biogenetic law in which embryonic development is a repetition of ancestral history, even if it is abbreviated in some

---

17 In the biogenetic law, embryos corresponded to earlier evolutionary forms and new developments were only added on at the end or old developments were truncated. Thus, a mammal recapitulates a fish during its embryonic development. Roux argues that different embryos differ throughout their development and mentions special adaptations for embryonic development. An example would be the development of a placenta by mammalian embryos for their nourishment in their mother's uterus. —Trans.

respects, is fundamentally wrong, just as wrong in principle as Newton's law of gravitation, Mariotte's law, and the law of fall are wrong in reality [insofar as real events never strictly follow them]. The first would only be correct if the masses of the bodies gravitating toward each other were united on both sides only in one point; the second would only be correct if the molecules themselves did not occupy any space; and, for the third, who has ever seen a stone thrown according to the law of fall or describing a real parabola? No one! And yet this will always be taught in school and rightly so. Analytical consideration is justified and compels us to do so. The law is correct if the component of air resistance is neglected but, in reality, air resistance always has an altering effect. The more powerful the effect of the perturbing component, the more the effects of other components will be obscured. In spite of this, however, the endeavor to understand the aggregate phenomena of change requires us to proceed analytically and to seek out the individual components and weigh them, one against another. Even when there is no wind, one does not observe parabolic movement of a thrown feather; yet, at the outset, the tendency of such a movement was communicated to the feather, and a pure parabola could be constructed out of the zigzagged line of fall, if one were able to deduct exactly the resistance of moving air.

The basic biogenetic law designates one formative component in the history of development, but the development of organisms is not just a straightforward formation of the complex from the simple, but detours occur and many backward steps. We only recall the well-known examples of the cleft gill and branchial arteries, which subsequently must close again, as well as the notochord and the superfluous, functionless structures of the pituitary hypophysis and pineal gland. With the rank of such an important, formative component, the "basic biogenetic law" will have its lasting justification. The magnitude of its recognizable effect, however, must be specially determined for each stage of the developmental history, for each organ, and for each class and species of animals.

Finally, allow me to say a little more theoretically about the degree of bequeathal, about the difference in the transfer of parental properties to eggs and to sperm. According to Karl Grobben and Moritz Nussbaum, in some animals the first reproductive products, the sex cells, are recognizable in the new individual before formation of the blastema. Because the sex cells separate themselves from their father so early before he has substantially differentiated, this indicates their high degree of self-sufficiency and proves that they receive the inheritance from their forefathers very early as instructions of *potentia,* before their father would be able to convert his special estate into individual formations.

However, despite this early separation from the father or mother, the sex cell remains dependent on and in contact with them, for it must feed, enlarge, and multiply. Moreover, it receives its food from the father through chemical material traffic, and by this route can also be influenced in its nature. Accordingly, it must be most probable that chemical differentiations, chemical alterations of the parents, are more easily transferred to the offspring than mere changes of form, such as the greater development of this or that group of muscles. Because we must ascribe the mental properties or temperaments to chemical alterations, not to morphological ones [?], the high degree of their heritability is understandable, as, in a similar manner, is the high degree of bequeathal of instincts and mental illness. So it is also conceivable that chemical alterations of other parts, such as more vigorous chemical constitutions of muscles or glands acquired through suitable food, are more easily transmitted to the child.

But whether parts with stronger material exchange, such as muscles, ganglion cells, and glands (the nutritional constituents of which are perhaps also present in the blood in greater quantities or diffuse more easily) transmit chemical alterations more easily than parts with weaker material exchange, such as the supporting substances, is not known. In any case, an analytical investigation would have to pay attention to this, in addition to the main observation of the difference in bequeathability of acquired formative and acquired qualitative characters.

The lesser bequeathal of properties acquired later in life relative to congenital properties already acquired in embryonic life could then be based partly on an ever-increasing self-sufficiency of the life of sex cells, which can prove themselves in selected properties despite the necessarily large food supply. On the other hand, there is the fact that altering influences in the embryo or youthful body do not remain just local-formal but, one might say, become more easily chemical. All molding, all shaping of the upper arm and its muscles is due to chemical conditions [?], although these muscles are not composed differently than the muscles of the thigh. Perhaps a formal change, brought about by an external influence on the embryo or the individual after birth, could more easily cause a chemical change and, as such, be more easily transferred to the semen. The ease of transferring chemical changes to the sexual products is best known from the transferability of infectious diseases, such as smallpox and syphilis, to the fetus or to the semen. As is well known, syphilis can be transmitted from the father to the child without the mother becoming ill (according to Wilhelm von Rosen, Jonathan Hutchinson, Ernst Fränkel, and others).

By tracing acquired changes in form to chemical changes and through the easier transferability of chemical changes to the sperm and egg in the chemical interchange of materials that takes place between the sex cells and the mother and father, the problem of bequeathal as such is eliminated, and the phenomenon becomes a more general problem, that of the production of form by chemical processes, which is the basis of all biology. Besides this problem, there remains the more special problem of the successive chemical changes in the chemical development of the egg, from which the successive formal development according to the first principle naturally derives.

The timing of bequeathal must still be addressed briefly but not in the hope that perhaps the primary, directly bequeathable characters should become recognizable earlier than the secondary ones that arise from them. For the contact in everything organic is very fine; and the primary is usually only a modicum of time and a modicum of space ahead of the secondary; so, unfortunately, for our observation they almost always appear simultaneous, and the reliable establishment of a causal connection can only be done experimentally, through altering one component. So it is not with this hope that we finally close this chapter (which is much too long for its necessarily meager content) on the temporal relations of bequeathal, but rather in order to obtain a more just judgment of the bequeathal of acquired characters.

Whoever regarded only the congenital characters as bequeathed could of course not identify functionally acquired adaptations of the parents as bequeathable; characters acquired in the twentieth year of the father's life do not immediately shift back to the embryonic period. As is well-known, this regression of acquired characters into embryonic life takes place only very slowly, whereas it goes without saying that embryonically acquired variations come to the fore immediately in the embryonic life of the off-spring. As a result of this slow regression, many generations must pass before a quality acquired in manhood appears in earlier youth. There-fore, with the changing occupations of people, it is very easy for an in-herited property, before it has become apparent, to be canceled again by a different way of life of the offspring so that its bequeathal is not at all noticeable.

These are probably the reasons why, for the recognizable bequeathal of so-called acquired changes, it is necessary for the transforming cause to act over many generations: on the one hand, to strengthen the character; on the other hand, to allow it to appear in earlier stages of life [?].

Furthermore, it seems to me to be a justified view that bequeathed ac-quired peculiarities of ancestors, instead of moving back to youth, can be suppressed by the changing influences of the outside world on the mal-leable, adaptable youth and become more and more prominent only at a more mature age,[18] once this interaction with the outside world has less effect. Darwin expresses this idea with an excellent example (without, however, developing the principle) by mentioning that with increasing age the handwriting of humans sometimes becomes more similar to that of the father. In accordance with this, I believe I have observed that family characters in men, especially intellectual ones, sometimes only develop and come to the fore in later years after they had been suppressed in youth by upbringing outside the family.

18 [Later I made an observation in myself that gives a different significance to this fact. When I answer a letter, my handwriting resembles the handwriting of the letter in front of me in size, steepness, and clarity of the letters, and in many other characteristics, without any special intention on my part; and when I write to people whose handwriting is well-known to me, the same happens to an even greater degree because I involuntarily have their handwriting in mind. I sometimes have to take special care to avoid this unintentional imitation.]

CHAPTER II

The Struggle of Parts in the Organism as a Cultivating Principle

Πόλεμος πατήρ πάντων.

—Heraclitus

A. Justification

The inscription of this chapter and book probably seems strange to some readers because it suggests that within animals, in which everything is so excellently ordered, in which the most diverse parts interlock so perfectly and work together as a highly accomplished whole, that a struggle takes place among the parts; that a conflict exists within the individual where everything proceeds according to immutable laws. And how could a whole endure if the parts were at odds with each other?

And yet it is so. Not everything proceeds peacefully in the organism, next to and with one another, not in health, even less in illness. In the latter case, the idea of internal disunity among the parts is commonly held, but the destructive effects of the same are daily before our eyes.

But how is the good, the lasting, to emerge out of strife and struggle? Perhaps someone may ask this question who has not been convinced, by the labor of the last few decades, of the general truth that all good comes only from struggle.

Heraclitus said, "Strife is the father of things."[1] The familiar conclusions that Empedocles, Darwin, and Wallace derived from this principle were discussed in the previous chapter. Just as the struggle among whole individuals leaves the best still standing, so may a struggle of parts have left, and continue to leave, the best, given an opportunity for such an interplay among the internal parts. Can a state not exist when its citizens everywhere compete with each other and only the most diligent attain general influence over events?

[The word "struggle" is used in the following mostly in a figurative sense, as often in Darwinist writings, in the sense that one is disadvantaged by the existence of the other, even if the two do not directly interact as happens in competition or contest. For this reason, instead of the term "struggle of parts," I later introduced the expressions "selection of parts" [*Theilauslese*] and "partial selection" [*Partialauslese*] to characterize the consequences of the "struggle."

Competitors do not need to know anything about each other for one to be disadvantaged by the other; for if the better competitor for a well-nourishing position had not existed, the inferior would have gained the position, and he and his offspring would not have starved.

Competition for space entails more direct interaction. Of course, one could, in this case, simply anticipate the other's space; but if the other attempts to enter the same delimited space, both must exert direct pressure upon each other. If two growing, expanding beings occupy a space that is already filled by them and bounded by fixed resistances, then expansion of one can only take place by compaction of the other, with direct active damage and cost to the other. The one that presses harder will gain more space than the one that is weaker. Such direct struggles are not solely confined to conscious beings with an intention to harm the other, but in essence also occur whenever two unconsciously expanding structures meet each other in a confined space. It is therefore impossible to see how Wilhelm Wundt could purport to find a "teleological reinterpretation"

---

1   The epigraph "Πόλεμος πατήρ πάντων." is usually translated as "War is the father of all things." In this paragraph, however, Roux translates the passage as "Der Streit ist der Vater der Dinge," or "Strife is the father of things." —Trans.

of "causal" connections in my application of the word "struggle" to such relations.

These types of struggle lead to a "selection" of the stronger, more lasting, or, as they say, to survival, to the more suitable being left behind. But the same result can also arise in some circumstances without any direct or indirect effect of structures on one another, simply by self-eradication or self-elimination, in that structures or processes that are not lasting under the given circumstances cease by themselves while others that are lasting persist, remain in existence, and must gradually accumulate in the world.

Both these types of "cultivating selection" occur within the organism and form the subject of the present chapter, with the formative effects of the qualities so cultivated set out in the later sections. This chapter and the whole book of which it forms part have been given the title *The Struggle of Parts in the Organism* according to the principle *a potiori fit denominatio*. But an alternative title *The Cultivating Selection of Parts in the Organism* might have been chosen based on the common result of these various processes, which perhaps would have been a better choice given the overriding number of those who read hastily.]

But is there an opportunity within the organism for such an interplay of parts to lead to the selection or cultivation of the best? This is the question on which everything hinges in the first instance.[2]

In order to answer this question it should first be mentioned that even in the highest organisms the centralization of the whole is not as complete as is often imagined. It is not the case that all parts can exist only in the organism to which they belong, only in their normal place of residence, only as standardized parts of a whole, and in full subjection to the whole.

Almost thirty years ago, Rudolf Virchow pointed out the self-sufficiency of cells and cited the ability of cells to be transplanted from one organism to another and from one place in the same organism to another. We are now able to transfer, from one individual to another, parts of the epidermis, whole pieces of skin with glands and hairs, also the completely detached periosteum, the cornea of the eye, and individual hairs. These transplanted

---

2 German *Instanz* has secondary meanings of a court or a trial. —Trans.

parts can then persist for a while or permanently in their new residence and possibly continue to grow. As is well known, this ability is more fully developed in plants, the organisms that gave propagation its name: buds, entire organ complexes, can be transferred; an excised twig can develop into a self-sufficient stem.

Virchow pronounced the following judgment: "If particular elements or groups of elements can be separated from federation with the human body, without ceasing to express and preserve their life characteristics, then it follows that this federation is not in the traditional sense unitary but rather is societal or, more precisely, communal (more social). Elements or groups of elements exit the body without the cooperative being destroyed; indeed, entry of new elements may even improve and strengthen the cooperative."[3]

Besides this demonstration that many parts are not absolutely dependent on the whole, a certain individual freedom of the parts is already expressed in embryonic development. The development of forms is not bequeathed as a standardization of the performance of each individual cell but rather as general norms of size, shape, structure, and performance of each organ so that wiggle room remains for the individual execution, for the construction from individual cells, within which events are mutually regulated.

We recognize this from the inequality of the parts of each organ. No liver cell exactly resembles another in size and shape, and yet all fit together to form a certain type of well-performing organ. It cannot possibly be determined from the outset by bequeathal that the hundredth, or hundred and first, liver cell has exactly this size and shape, slightly different from all others, and connects with the previously formed and subsequently formed cells at precisely this angle that is slightly different for each; rather, the following cell attaches itself to the preceding cell according to its individuality, constrained only [apart from adjusting itself to the shape of spaces between capillaries] by needs inherent in the cells' bequeathed "quality" for some contact with capillaries, but otherwise free in its associations with neighboring cells, etc.

---

3   (Virchow 1880, 186).

Embryonic happenings apparently take place like the undertaking of a contracted work, for example, a construction, for which material, size, shape, internal furnishings are standardized merely insofar as they are determined by the intended use, hence by the function of the house. But much of the individual execution is left free to the contractor and his assistants: for example, the laying of the individual stones, especially irregular natural stones whose placement together can only be done in a particular way if they are to perform the required function. So one stone after another is inserted, and the following one is adapted to the previous one in position, size, and shape, or possibly the other way around; if the next one is large enough, it forces the previous one to adapt to it.

But through all this no struggle develops in our sense, no preference in the interplay of parts leading to the cultivation of the more suitable. We can only reach this conclusion when we take into full consideration the "vital properties" of the organic, and especially the properties of growth and proliferation.

In organic beings, the building stones are not ready-made and then simply assembled one after the other. Instead, subsequent stones are always the products, or descendants, of earlier stones. If the existing cells of a youthful or still embryonic organism are not all alike, but instead there is a single one that is favored by some special bequeathable quality and is able to produce more progeny than another, then the one that produces more progeny will come to comprise a greater proportion of the whole organism; and when its descendants have inherited the favorable quality from it, their already larger number will by further proliferation take a leading role in building up the whole.

If the individual is already grown, and we are concerned only with physiological regeneration, then exactly the same thing can occur; for as soon as a cell is dying, *ceteris paribus,* whichever of its neighbors is more powerful according to its nature will tend to proliferate the most and replace the departed cells. Because its progeny will again be more powerful when such opportunities are repeated, they will gradually penetrate into ever wider circles.

But, as we have assumed, such a struggle is only possible if the parts are not perfectly equal to one another in qualities that can be bequeathed to their progeny, and are therefore unable to keep one another constantly in balance. With absolute equality of all parts that function alike, the share of each in the structure of the organism or in its regeneration must be the same, and only "external" favorable factors, such as a more favorable position in relation to a blood vessel, could produce favorable treatment, which, however, would only be slight and temporary because it could not be transferred to descendants. But if it were transferred to descendants, this would demonstrate that the favorable treatment was based on the nature of the maternal cell—that is, it was an internal, not an external, favor.

"Qualitative" inequality of parts therefore must form the basis of the "cultivating" struggle of parts. From this inequality, a contest arises by itself because of growth and, as we shall immediately add, also simply because of material exchange. Because all parts consume themselves in material exchange, they will have to nourish themselves for preservation and production; and thereby parts that, due to the available food or for some other reason, are of inferior quality—that is, are unable to regenerate themselves as quickly and as completely—will soon be at a considerable disadvantage compared with the more favorably disposed ones.

But the presupposition of the qualitative inequality of the parts from the beginning, is it present? Is it not an arbitrary assumption? Nowadays only a layperson who might glance at this treatise will ask such a question, now that we have become accustomed to paying attention to all the differences, even those that are subtle. Every naturalist knows that the same being never survives unchanged for a long time, never recurs in exactly the same way, that everything, the inorganic as well as the organic, is constantly changing.

How difficult it is, and what special precautions are required, to keep relatively simple things equal, to produce a uniform mixture of glass for the objective lens of a larger astronomical telescope. How dearly we must pay for each uniformity in all products of our industry, be it uniform substances or colors, or an even division or thickness or surface—in short, every uniformity in a larger space or in repetition of many objects—because it is so difficult to keep something constant, because everything, even metal machines, changes constantly, whether by heat or by wear and tear or by something else. Nothing can be kept absolutely constant because everything constantly changes, because everything influences everything else. Always the living forces fill the space, be it in the form of mass motion or molecular motion as heat, light, or electricity. Always the living forces change tension forces and materials. Nothing is isolated in the world, least of all the organism, which must continually absorb and convert substances from the outside. The more complicated the event, the more difficult it is to maintain constancy: if two crystals are never completely alike in all their properties, how much less alike must two organisms be?

[Morphological assimilation, the greatest riddle of the basic organic process, is of admirable perfection. In the face of constant change in external circumstances, it must therefore be singularly safeguarded by mechanisms of self-regulation. But there is no such thing as complete perfection; therefore, it too must give rise to alterations, to variations in the materials within cells and among cells of the same tissue. And if such variations are minimal in germ cells, they will be considerably greater in the cells of the developed individual.]

No two things are identical in form and quality—not the young of a litter, not parts of an organ, not cells of the same tissue. This is manifestly useful because not all cells are at the same stage of their life at the same time; otherwise, all cells would undergo physiological death at the same time, and the organism would be destroyed by the failure of the organ in question.

It is true that the organism [apparently through outwardly complete self-regulatory mechanisms in all formative processes] is now regulated so that it maintains itself approximately constant despite change in external conditions and the infinite complexity of its own interior; but this *constancy* is only approximate, observed only fleetingly; and steady changes, as Darwin has taught us, can accumulate to a considerable degree. Variability is even greater at lower levels of organic life; and it must have been greater still in earlier times, before these organisms had developed regulatory abilities and a particular balance with their surroundings.

So every seed animalcule and every egg is already differentiated from the other; and, because the essence of development is to evolve the heterogeneous from the more homogeneous, the complex from the simple, it is particularly obvious that these formations of different qualities and forms will be somewhat deflected by altering external influences, so new differences are brought about among the parts of the organism.

Because of these inequalities, which are continually brought about not only in the whole but also in the parts through the change of conditions, it was impossible from the beginning that laws of formation could have developed that standardized individual occurrences down to the last cell and last living molecule. Such laws could never have led to the building up of an organism with the constant change in circumstances, such as a field commander could not win a battle who, instead of broad orders to his generals about the formation and deployment of their troops, at the outset issued special orders down to the deeds of the lieutenant or individual soldier: for the accomplishments of all must be continuously adapted to changing circumstances, all the more at the small local scale because its circumstances are more easily changed than those that happen on a large scale. The individual cells must always be able to adapt to one another and to novel conditions brought about by changing influences.

The cultivating struggle between the living parts, evoked by the differences among them, will therefore have begun with the first origin of life and never ceased since then; and it is natural that the most general properties were cultivated first, so that the first beginning of what we will develop in the following is to be sought in part already at the time of the origin of the organic. And it is just as obvious that in times of *greater variability* the struggle of parts must be correspondingly more fierce and of greater importance than in times of approximate *constancy of the species*.

But on how long this took—or, physiologically speaking, how many generations were necessary for the development of the properties to be discussed—we can no more say anything even approximately correct, than we can say about the magnitude of variations that once occurred in earlier times nor about the vigor of earlier life processes.

[Before we proceed to specifics, some precursors in the understanding of the struggle of parts must be reported.

First, George John Romanes claimed priority for Herbert Spencer by pointing out that he had already made many comparisons between an individual's organs and processes within a state as a social organism. We find these comparisons in the first volume of Spencer's *Principles of Biology* where he seeks 'to obtain the key for the laws of adaptive changes in organisms from those in society.' But his comparisons refer only to the functional correlations of various organs with one another—that is, to functional adaptation—as well as to the competition of the various organs for food, but not to a struggle of similar parts, which, as we have seen, leads to the selection of the best and consequently forms the basis of the discussions in this book because it gives the basis for a new explanation for functional adaptation that is fundamentally different from that of Spencer, while the older explanation accepted by Spencer is proven by me to be incorrect.

Furthermore, Charles Darwin, in the letter cited previously, in which he had occasion to mention a review of my book, wrote, "I believe that G. H. Lewes implied the same fundamental idea, namely that within every organism there is a struggle between the organic molecules, the cells and the organs." I have, however, not found such a position adopted by George Henry Lewes; and even Romanes, to whom this letter was addressed, does not indicate such a thing in his review, so this matter appears doubtful at this time.[4]

On the other hand, as I later discovered to my chagrin, Ernst Haeckel had priority to a considerable extent. In his work on the *Perigenesis of the Plastidule or the Wave Generation of the Particles of Life* the following statement can be found: "By so doing Lamarck's doctrine of the bequeathal of modifications (which is the most important presupposition of Darwin's theory of selection) can be transferred from the great multicellular animals and plants, which appears palpably before our eyes, to the plastids (cytodes[5] and cells) and from these in turn to the plastidules that make up the plastids, we naturally assert for the latter the consequences established from the theory of selection for the former. Obviously there is 'the struggle for existence among the molecules' in the truest sense and above all among the active plastidules, which Pfaundler first illuminated in 1870. Those plastidules that best adapt to their own conditions of existence, those that take up the liquid nourishment that penetrates from the outside most easily and most readily carry out the resulting rearrangement of their atoms, will naturally exercise the strongest assimilation and thus predominate in the reproduction of the plastids."

Haeckel thus transfers the struggle of individuals to the nourishing parts of the organism (in general) and deduces from it that those cell parts that can best take up and assimilate food can multiply most rapidly and

---

4   George Henry Lewes discusses internal struggles in *The Physical Basis of Mind* (1877), 101–111. —Trans.

5   Cytode was Haeckel's term for a unicellular non-nucleated mass of protoplasm. —Trans.

therefore acquire a greater share in the structure of the individual than others less qualified in this direction (in special).[6]

In a lecture, delivered in Prague in 1879 on "competition in nature," William Preyer viewed the struggle of parts in the organism from a different angle than in Haeckel's short description.[7] He said that "the law of competition permits an application to all living things, in the broadest sense, that has received little or no attention. Not only do humans, animals, and plants compete with one another for what is necessary for the preservation and beautification of life, but, at least with regard to preservation and expansion, so also do the growing parts of which the individual organisms consists and the natural groups in which plants, animals, and humans live together."

After mentioning the division of the egg into cells and the differentiation of the cells into tissues, Preyer continues, "All these tissues grow, and it is clear that they must mutually affect one another as they grow, because all growth requires space. In the egg, however, the space is limited, and although in many animals the egg itself grows as the animal grows, here too there is a limit which will soon be reached. So it happens that even in the infant, the developing, and the grown being, almost all parts with all functions compete with each other. A normal continuance is only possible through the most uniform possible effectiveness of all parts, through com-

---

6 [In 1879, I informed Herr Professor Haeckel in a letter of my intention to write a treatise on the struggle of the parts in the organism and its selective and formative effects and asked him for his opinion on this project. He approved of the same and recommended adding to the main title "The Struggle of the Parts," the addition "as a cultivating principle in the cell-state [*Zellenstaat*]." For reasons of simplicity, I omitted this very significant addition at the time, much to my own detriment, because some readers did not grasp this main point and only took from the book (or the title page?) that I meant "struggle of the parts" to refer to "correlations in the organism." Because Herr Haeckel's answer did not inform me of his priority, and because there was no mention of the struggle of the parts in the relevant main writings of Haeckel, in the *General Morphology of Organisms* or in the *History of Creation*, I believed that this idea was unfamiliar to him. The same error was committed by his colleague from Jena, Professor William Preyer, who, like me, first published a paper in 1879 in which the struggle of the parts was briefly referred to, without considering Haeckel's priority.]

7 Preyer's lecture, "Die Concurrenz in der Natur" (Competition in nature) was delivered in Prague on December 3, 1878. Roux's footnote at this point cites the publication of the essay in Breslau in the February 1879 issue of *Nord und Süd* but states that he quotes from the "second printing" of 1882. There were in fact three printings (Preyer 1879, 1880, 1882). The date of the lecture is given in Preyer (1880). —Trans.

promises, of which I have already spoken. In fact, one-sided, exaggerated formation and activity of a tissue or some kind of organ always takes revenge.

"For just as cells compete with one another and tissues composed of cells compete with one another, so do organs and organ complexes composed of tissues compete with one another in every organism. And here lies the cause of the limited size of each part. The liver, the lungs, the eye, cannot grow beyond a certain extent because of competition with other organs. This necessarily results in the limitation of the capabilities of the parts, and thus of the whole. Even the experience of daily life shows that muscles weaken in those who think and study a lot, whereas those who only work mechanically with the hand seldom solve difficult problems of thinking; the blind feel and hear very well whereas the deaf often see very well. Almost all animals provide examples of these consequences of the competition of organs for their functions, which allows preponderance of one only to come about at the expense of others."

In these ideas, as in the earlier arguments of Herbert Spencer, competition of *dissimilar* parts is represented, which, as we shall see, leads to a certain harmony of the parts of the whole organism. The main subject of my discussion will be competition among *similar* parts, which, as we shall see, leads to the selection of qualities that increase the lastingness of the individual and to a new and sufficient explanation of functional adaptation. This subject will only be considered in general, and its special consequences will not be discussed.

I first received news of Preyer's publication through the separate version of the same from 1882. In the spring of the same year as the first printing of Preyer's lecture appeared, I concluded a treatise[8] in which I already used the struggle for food and space of the same-functioning parts of the organism to deduce purposive organic formations and moreover already introduced the new "principle of the victory of those who function more strongly in the specific manner," on which my explanation of functional adaptation is based. In that treatise, this principle was already extended to the "struggle of ultimate cellular particles" by "which in material exchange, only the processes that form the specific function that

---

8 Wilhelm Roux, "Über die Bedeutung der Ablenkung des Arterienstammes bei der Astabgabe," *Jenaische Zeitschrift für Naturwissenschaft* 13 (1879): 321–337. —Trans.

are constantly stimulated and thus become strengthened by continual functioning were able to regenerate themselves lastingly with materials at the expense of the less stimulated, less specific processes." And I already drew the consequences of this struggle for the origin of the functional shape of cells; almost all of the consequences that are subsequently presented in detail in this book.[9]]

B. Kinds and Achievements of the Struggle of Parts

If, after this general justification, we proceed to the investigation of the special kinds and achievements of the struggle of parts, then it must necessarily be broken down into as many sub-instances as there are independently varying life units—that is, into a struggle of living cell particles, of cells, of tissues, and of organs, each unit fighting only with its own kind. For a struggle between members of different units—for example, a plasson molecule[10] with a cell, or a cell with an organ—would be like a summation of differentials of different orders. Only when the property of a particle of the lower order has increased by spreading to an individuality of a higher order, only when the second-order differential has been integrated into a first-order differential, can the struggle with another individual of this higher order begin.

1. *The Struggle of Living Molecules*

For the sake of brevity,[11] I have chosen the above designation for the struggle of all kinds of the ultimate living cell parts, corresponding to Ernst Haeckel's

---

9 [Mr. W. Preyer does not have any priority in this matter that was laid down in print. But since, as already mentioned, I was fortunate at the beginning of the seventies to attend lectures given by Professors Haeckel and Preyer, in listening to these lectures it is reasonable to assume that one or both gentlemen, be it directly or indirectly, planted the germ of thought for this work in me, which then arose a few years later and developed into the present book. I also owe to those exceedingly stimulating lectures the introduction to the circle of ideas in which the whole of the text moves.]

10 *Plasson* was the protoplasmic substance of a cell. —Trans.

11 [Some authors read this work so cursorily that they believed and referred to the following discussions as concerning physical molecules, although they deal with material exchange, growth, and other functions of the parts. Therefore, I have now added the epithet

plasson molecules, Louis Elsberg's plastidules, Herbert Spencer's "physiological units" or the smallest units of organic process.[12]

If, according to our presupposition and apart from the cell's division into cell body and cell nucleus, the living (at least, the assimilating, growing, and multiplying) parts of a cell are not exactly alike, not exactly the same among themselves, perhaps minimally different in individuals with no variety in the tissue to which the cell belongs, somewhat more different from each other in individuals with new varieties, *then* these differently qualified substances of the same cell [each of them capable of assimilation and transmitting its properties to its offspring] must necessarily behave differently under different circumstances.

Let us assume that two different qualities of such living substances were originally present in the cell in equal quantities, and let us consider their behavior in *material exchange* during the period of growth. Then, in the replacement of what was consumed in material exchange, the substance that is endowed with stronger affinities and that more strongly assimilates will regenerate faster than the substance that is less endowed with these properties. The former will therefore, *ceteris paribus,* unfold spatially more extensively than the other over the same time period and thus take away its place. At the next repetition of this process the weaker party, which now already occupies less space, would again be unable to regenerate itself as rapidly and would again suffer a percentage loss of space; with longer duration, it will be pushed back more and more and finally dwindle away, and the time that this takes will depend solely on the size of the difference in the affinity of the substances that are in all other respects equally viable.

"living" throughout to "molecules." This term, "living molecule," means at the very least the smallest particle still capable of assimilation and growth, which is isoplasson; but mostly my discussion addresses the increase in number of particles as well as special functions. Therefore, the living molecules in question are of course the smallest structures that also have these functions in addition to those mentioned earlier.]

12 [These have lately been given more consideration and are called plasomes by Julius Wiesner, biophores by August Weismann, and pangenes by Hugo de Vries. I have classified these ultimate elementary organisms, or rather bare elementary organs, according to the functional differences to be assumed analytically, as isoplassonts, autokineonts, automerizonts, and idioplassonts.]

In every cell then, *ceteris paribus,* the substance that under the given circumstances (of blood composition, diffusibility, and the like) regenerates more rapidly and grows more rapidly will be preserved, and those with other qualities will be suppressed.

In this case, a struggle for space takes place as a result of unequal growth; for if space were unlimited, the weaker substance would be able to compensate, more or less, for its disadvantage by a longer duration of regeneration, provided that consumption is not continuous and equally strong, but instead pauses occur during which regeneration is stronger than consumption.

If the differences between two substances are such that they consume their own substance at unequal rates, this will *ceteris paribus* have a detrimental influence on the more rapidly consumed substance if the average consumption is large, and the more slowly consumed substance will dominate; because it consumes itself more slowly, but according to our assumption regenerates just as quickly as the one that consumes more quickly, it will increasingly predominate spatially and thus occupy the place alone, while the other disappears through self-elimination.

[In this example, we reckon with a difference in regeneration and at the same time with a corresponding difference in growth of the two substances. Such a connection is highly probable because the physiological regeneration of cell parts and their growth are probably both based on the same basic process of assimilation. But this is not the easiest case.

In order to discuss this easiest case, let us assume that the two substances that regenerate themselves unequally quickly are unable to grow but simply consume new materials to replace mass that is lost and that the supply of new materials is less than the combined rate at which the two substances lose mass. Then the substance that regenerates its mass more slowly will be unable to completely replace what has been lost, while the substance that regenerates more quickly can replace lost substance with new materials as they become available. So the mass of the first substance will decrease, losing more and more mass the longer this period of consumption lasts, until it finally disappears altogether, whereas the other substance maintains its original mass. So the substance regenerating itself more slowly will disappear by self-elimination, simply due to its insufficient lastingness, without causing any disadvantage to the other substance, and only that other substance will remain.

If we assume that, before the less-favored substance has completely disappeared, the amount of new materials required to sustain its diminished mass decreases sufficiently that it can completely regenerate this lesser mass, then both substances will remain in the amounts present at this moment, since we assumed that they are merely replacing materials as they are used up to restore their present organic substance to its steady state (and to store working materials) but not to increase their organic mass by growth. If both also are able to grow in the same specific measure—that is to say, the same mass of each grows the same amount in the same time—then the original proportions of their masses will remain pretty much unchanged after growth as long as consumption is sufficiently small relative to the supply of new materials; only the less favorable substance will proportionally lag behind to some degree because its growth is somewhat delayed as a result of its slower regeneration.]

That the struggle must be over space we shall show in detail on the basis of existing direct observations, when considering the higher spatial unit within which the above struggle of living substances takes place, in the struggle of cells. [The law of impenetrability must, of course, apply to living as well as inorganic bodies.] In any case, the struggle for space "within" the organism, where everything is connected to one another in a spatial unity and presses against each other, must be much more violent than the struggle among persons, among free individuals themselves; [for every part that grows more strongly than its surroundings must passively stretch the adjacent parts or cause them to shrink by compression.] The fact that spatial restraint is really able to inhibit the development of cells is shown, for example, from the flattening of epithelial cells against one another, and from the change in their shape immediately after the spatial restraint is removed. Thus, after a loss of epithelial cells, the tall and narrow columnar epithelium of the trachea becomes short and broad (flattened) and proliferates.

Apart from that though, even if the size of a cell is not limited by external restraints, its possible viable size is limited according to its composition, the nutritional conditions, the flexibility of its protoplasm, and other factors. This is likely a consequence of the limited radius and speed of diffusion, for if the cell were to expand beyond this limit, its interior must atrophy because of lack of opportunity for regeneration. [This kind of spatial restraint can mean that even in free cells—in other words, those whose spatial expansion is not constrained from the outside—a substance that has been reduced in its quantity within a cell cannot, or cannot entirely, compensate for its loss relative to the other substance, even if it continues to grow for a longer time than the other substance, *ceteris paribus.*]

If the substances are so different that one can regenerate itself more completely with the available quality of nutritional material, then this more favorably placed substance will become stronger, will proportionately predominate, and, with its correspondingly greater growth, will displace the other in the struggle for space.

If there is a change in the nutrition of cells, in the composition of the blood, then according to these changes other chemical qualities will be enabled to dominate and displace the earlier ones. [These will be the qualities of living substances that can best regenerate and proliferate with the new materials while the other qualities eliminate themselves in the ways just discussed or are annihilated in the struggle for space. Thus, through internal recultivation (*Umzüchtung*) within cells, cells become adapted to new modes of nutrition, provided that the cell can survive this change.]

If food shortage persists, no fighting for space can occur, but only those associations will remain that, *ceteris paribus,* use the least amount of material for replacement while those that require more will starve and disappear by self-elimination. [The more unfavorable substances are disadvantaged by the existence of the more favorable ones, possibly even brought to starvation by them, in that the latter remove part of the, already scanty, food, especially when one substance has stronger affinities for food than the other. In states of hunger, there is an indirect struggle for food, in which the cell (or whole organism) is recultivated as a "saving machine" that works with minimal consumption and highest affinity for food.]

Among the living substances within a cell, if there were a substance with the property that its ability to absorb and assimilate material for its regeneration from the immediate environment increased as consumption increased—to speak in the language of whole individuals, if there were a substance whose appetite and rate of regeneration increased with its need for food—while other substances did not possess this property, but always attracted and assimilated food at the same rate, corresponding to the average rate of consumption, then this substance must triumph over the other substances if increased consumption lasted for a prolonged period, for this self-regulating substance would regenerate itself more completely than the other substances. (This important case has already been mentioned occasionally.)

Were the composition of a variety of living substance such that in material exchange its assimilation exceeded its decomposition while other substances lacked this property, then overcompensation of the consumed (in other words, growth) must occur, and this important quality must win sole control over all other qualities, as it is known to have done. We know of no organism or cell that does not have this property of growth, overcompensation for what has been consumed, in one stage of its life; and it demonstrates how, without these properties, life could not have spread at all, that the life processes would always have had to remain limited to the dimensions in which they originally arose. [Without growth, a destructive external influence could easily have destroyed the life substance *in toto,* but this could have happened much less easily with the greater dispersion caused by growth. Furthermore, the greater mass of the active substance itself already represents a superiority over that which is only capable of lesser dispersion.

A mass simply growing faster than any other will likewise gain these advantages, albeit to a lesser extent; for if two substances are present in the same quantity but grow at different rates, soon, as a result of the geometric progression of increase in growth, the more rapidly increasing quantity will greatly predominate; and this will be the case, in greater and greater proportions, the longer this unequal growth persists. For example, if the rates of growth of substances a and b, initially present in the same amount, are as 3:4, then in the time it takes a to have tripled, b will have quadrupled; in another such period, the amounts become $3 \times 3a$ and $4 \times 4b$; then in the next period, $3 \times 3 \times 3a$ and $4 \times 4 \times 4b$; then again $3 \times 3 \times 3 \times 3a = 81a$ and $4 \times 4 \times 4 \times 4b = 256b$; and so on in geometric progression. But if the space within which these growing structures can expand is constrained, then, as we have seen, a direct struggle will soon take place for space. In this struggle, the life substance that is more able to withstand pressure will triumph. Thus, substances of high tone are cultivated.]

These, then, are all properties that, as a result of material exchange and growth, must attain dominance within the cell along the track of the struggle of parts for food and for space [and partly by self-elimination of the less lasting], as soon as traces of these qualities had appeared as a "hereditary" variation within cells. So this is true then to the extent that the first pre-condition holds of the composition of the cell not being homogenous, but instead the same variability obtains for the parts of the cell as for the individual.

But who would like to dispute this probability, who would like to assume that when the organic came into being the substances were entirely of the same kind in the directions mentioned; and that when the infinitely many qualities of the organic, which we find in the various organs of the various classes, genera, and species of the animal kingdom, came into being, that they always appeared by themselves in a perfectly homogeneous manner, so a struggle could not occur among their parts?

Of course, these qualities do not need to be developed everywhere and at the same time; and it may well happen that a living substance, by virtue of excellence in one of these properties, in spite of a defect in other directions, retains dominance in the cell as long as the circumstances do not change to give other, in many ways more favorable, qualities a significant advantage over the defect.

Anyone who knows the precision of the relevant processes (insofar as physiology has revealed them) will say the deduction of these properties is highly probable, rather than arbitrary and drawing on putative eventualities that have never occurred in organisms. (These deduced properties are: the least consumption, the fastest and most complete regeneration with the smallest amount of material, the development of the most-lasting qualities given the available food, self-regulation by increasing hunger, and assimilation with increased need for food.) [This is particularly supported by George Katzenstein's finding that up to 35 percent of food can be exploited for walking performance, but the comparable efficiency of our best machines does not exceed 25 percent.] But if the organisms really have these favorable properties, if such material variations were even possible and occurred, then they must have developed by themselves in the way described through the struggle of parts without participation of the struggle for existence among individuals.

But parts do not just live quietly for themselves in their material exchange but are often struck by external influences, *by stimuli*,[13] and their processes are thereby *influenced,* perhaps accelerated. If the cell is composed of different substances, each stimulus must have different consequences for different substances. For our purposes, however, their behavior comes into consideration only for stimuli that "often" have an effect, that are often repeated during life, because they alone will be able to bring about "lasting" changes, to gradually cultivate certain qualities in the cells. The mode of action of the encroachments of these agents, these living forces, can be very different.

If, through accidental variation, one of the different cell substances is such that, *ceteris paribus,* it is used up less quickly during the transformation caused by the stimulus than the others under the same stimulus, then it will gain sole control in the cell easily, as was shown above for the substances less rapidly consumed in material exchange. In the same way, a substance whose affinity for food and ability to assimilate food is increased by the stimulus will eventually triumph and finally remain alone, for it has an essential advantage in its increase over others that are less favorably influenced by the stimulus.

Now if there were also organic processes that were not simply strengthened in their regeneration by the stimulus but, in responding to the stimulus, were strengthened to the point of overcompensation for what was used up, then these processes would have brought the property of growth into dependence on the stimulus. According to the above, such processes if left to themselves must achieve victory over all others and achieve sole control even sooner; for they would, *ceteris paribus,* proliferate even more

---

13 [Some authors (such as Carl Weigert and Ludwig Edinger) regard the "stimulus" as something mysterious or mystical and therefore try to limit its part in organic events as much as possible, even to eliminate it completely. Although the special stimulus is in some cases unknown to us, there is nothing mysterious in the principle of the stimulus in itself. The simplest definition given here as an "external influence," thus as a supply or withdrawal of living force (kinetic energy) contains absolutely nothing mysterious. The latter, on the other hand, mostly lies in the nature of the reaction of the organic to the stimulus, insofar as this triggers complicated mechanisms, for example growth or greater resilience. But the latter two are not necessarily included in the present case, as can be seen, and are not accounted for by the stimulus but by the stimulable.]

than the other substances and therefore, from the beginning, push them back more and more proportionately, finally to nothing, given the limited space for expansion permitted by the diffusion conditions.

The assumption was made here that living substances have appeared and dominated whose viability is increased, especially their assimilation, by the supply of stimuli. That is, the stimulus was assumed to have a trophic, nutrition-enhancing effect. Perhaps at first sight this assumption will appear completely arbitrary to some people, at least for animal processes. However, the corresponding behavior is known to be the basic condition for the existence of plants because their assimilation depends entirely on sunlight and warmth, and Carl Wilhelm Siemens has recently demonstrated a similar influence of electric lighting through the unusually rapid development and fructification of plants in such lighting [while, on the other hand, electric lighting has been observed to inhibit elongation in plants].

But stimulatory effects, especially of light, are also known for animals. Jules-Auguste Béclard found that eggs of the flesh-fly *Musca carnaria* develop most rapidly in violet light, and successively more slowly in blue, red, white, and green light. Émile Yung similarly found violet to be most favorable, then blue, yellow, and white, while red and green were directly injurious for development of the edible frog *Rana esculenta* and the common frog *Rana temporaria;* and darkness retarded the development of these animals. (A different effect of light on the development of animals, a kind of heliotropism, had already been demonstrated in 1870 by Leopold Auerbach. Pigment always formed or accumulated on the illuminated side when he illuminated frog eggs from below. Samuel Leopold Schenk has recently observed the same thing, particularly strongly in blue light, less strongly in yellow.)

[The strong stimulating influence of light on the material exchange of animals and humans, reported by Jacob Moleschott, Carl Voit, and Max Pettenkofer, Karl Speck, and others, should also be pointed out.

The supply of warmth is one of the most important conditions of animal and vegetable development; the germination and spawning periods of these living things are based on it. And we are able to slow down or accelerate their development by withholding or withdrawing this agent.

Eduard Pflüger bases his law of the proportionality of life processes to temperature on corresponding observations.

Mechanical shock is also important. Fritz Müller reports on larvae of *Paltostoma torrentium:* "Like some other animals that live in strongly agitated water (in sea surf, waterfalls, or rapids), these larvae and pupae soon die when they are brought into calm water." Because these larvae attach themselves to fixed objects by means of their suction cups, they are not found in the water foam; and I conclude from this that it is not the abundant amount of air found there that is their need for life, but the shock as such, the supply of this form of living power.

In organisms that are accustomed to develop at rest, on the other hand, shock has a disruptive, even destructive effect on the capacity for development, as Camille Dareste has reported for hens' eggs that were transported long distances in a wagon. Eduard Pflüger and Dietrich Barfurth found that the metamorphosis of amphibians was delayed by mechanical tremors.]

Ewald Hering also asserted trophic effects of stimuli for other animal parts in his theory of the sensory nerve function, assuming that in the sense organs particular stimuli increase assimilation whereas others increase dissimilation or decomposition. Contemporary physiology also assumes this effect, in principle, in its doctrine of trophic nerves but ties the supply of such stimuli to special nerve tracts. [Even if, as will be explained below, we generally doubt this last assertion, the principle of increasing assimilation through stimuli is nevertheless assumed and thus recognized by it, and the electrotherapeutic experiences of Archibald Reid, Robert Froriep, Guillaume Duchenne, George Beard, and Alfonso Rockwell, and others with muscular and nervous diseases are based especially on this principle.]

But once such variations of cell substance appear—the life force of which is increased by the supply of different stimuli or by just one particular stimulus—that variation which absorbed the stimulus more easily than the others must always gain the victory and gradually sole existence, *ceteris paribus.* As a result of this property, its vitality is strengthened, and it must thus proliferate more. If the size of the stimulus was limited, a kind of indirect struggle for stimulus, and victory through stimulus withdrawal, must arise as in the case of the competition for food because the more easily excitable substance would absorb relatively more stimulus and thereby become capable of greater unfolding and consequent greater spatial spread.

If such stimuli now act continuously then with further variations, with the increasing perfection of the adaptation of substances to stimuli through ever new struggle selection in cells, the path is taken to an ultimate final stage, in which processes must remain behind that were in the highest degree able to absorb the stimulus and be strengthened by it but are no longer able to keep themselves alive without the stimulus. In other words, these are substances that must consume themselves without regeneration if the stimuli are absent because *these stimuli had become indispensable life stimuli for them.*

We shall see later how important such a high degree of adaptation had to be for the perfection and formation of organisms and that we are entitled to ascribe such properties to some of our cells.

Furthermore, if stimuli strengthen vital processes, then different stimuli must strengthen different living qualities.[14] Adaptation to the different natures of the stimuli must have occurred, and it must have been increased by the struggle of parts with new variations whenever the same stimulus acted on a cell.

If, on the other hand, different stimuli act alternately, but repeatedly, on the same cell, and if the cell contains a substance that is strengthened by both (or more) stimuli, then this substance would eventually gain dominance, depending on the kind of pauses and the relation of the nature of the stimuli to each other. But this versatility of a single substance would, however, only rarely be maintained, for such a substance cannot comply with each stimulus to the same extent, cannot absorb each stimulus so easily and implement it so perfectly, as a substance especially adapted for a single stimulus. This is due, first, to the fact that every change in the substance's constitution, which arises for first the one then the other stimulus, must always necessarily be connected with a loss of force because there has to be a new rearrangement of the molecules. And second, a substance can never be so perfectly adapted to two different stimuli, and there-

---

14   The first edition has "different chemical qualities." —Trans.

fore also not so strongly strengthened, as can be a particular substance for each individual stimulus.

But when two or more different qualities of living substance have appeared in the same cell, each of which is particularly strengthened by a different stimulus, then when these stimuli alternate, each will be able to permanently occupy a certain volume in the cell, depending on the magnitude of the stimulus in question, in proportion to the magnitude and restorative capacities of the other stimuli for the other substances. The cell will be able to remain composed of very different substances, which correspond to the different nature and magnitude, as well as the intensity and frequency, of the stimuli in question as we see developed in protozoa.

The struggle of parts and self-elimination will thus at the same time be compelling principles for ever more extensive differentiation, for ever more perfect special adaptation to the stimuli, insofar as stimuli are able to strengthen the life processes. That stimuli indeed have these effects we shall explain in detail in the next chapter.

If, with the external circumstances of an organ, the stimulus to which it was once adapted changes, the old qualities will not only be subject to atrophy, declining as a result of lack of life stimulus, but also the substances strengthened by the new stimulus [as soon as such have occurred, even if only in small quantities but with a bequeathable quality] will win victory, directly affecting the others in the manner described and accelerating their regression and transformation.

Just as the new cell is capable [with sufficient variation of its bequeathable quantities] to adapt itself to different stimuli by increase of different constituents, so differently constituted cells can adapt to the same stimulus in different ways; for, depending on the particular nature of the cells, the same stimulus in each of them can most strongly strengthen different substances in their assimilation.

All these properties, which are extremely useful and purposive for the preservation of whole individuals, are thus preserved as soon as they have emerged in traces and expand in the cell with the suppression of less "lasting" ones. And as soon as substances appear by new variations that have these properties to an even higher degree, these will conquer the earlier ones; and thus, on and on, selection from the variations will increase the degree of quality.

All of this happens without the struggle of individuals—indeed, possibly against it because it must appear questionable whether selection by the "struggle of individuals," even if it concerns very useful things, could cultivate something that could not survive victorious in the struggle of living molecules.

The speed with which new variation spreads within a cell cannot, of course, be judged with certainty. But it can be assumed that, with the steady persistence of material exchange and therefore also of the struggle, this can lead to the sole existence of the more favorable property in the life of one cell or even more quickly, unless new, more-lasting variations appear in the meantime [which now push back the formerly favored substance]. Possibly, the perfect balancing out first occurs in the cell's offspring.

From the properties that have been preserved and disseminated in this way, which have proven to be the strongest and most *lasting* of their kind, selection in the "struggle of individuals" will choose for truly lasting preservation only from those that prove useful "for the whole individual." Thus, for example, from all the living materials that are increased in their nutritional ability by a stimulus and that use the least amount in implementing their reaction, the one that contracts most strongly and fastest is selected at one point in the organism; at another place, it is the one that best uses the stimulus to attract and transform substances that are then deposited; in a third place, it is the one that uses the stimulus the least, but rather lets it go farther and guides it. Thus, the struggle of individuals will, by separate selection, develop particularly qualified structures from the most high-performing muscular, glandular, and neural substances cultivated in the struggle of parts.

If the stimulus is very specific, then the range of eligible processes will be somewhat narrower—for example, for the effect of light. But even so, the choice is among very different kinds of reaction, as we can see from the fact that at the same time qualities have been obtained in the same organism that react to light with pigment formation, whereas the photoreceptor cells are excited as strongly as possible by the agitation of light without consuming it but rather pass it on as strongly as possible. And it is necessary, as we have shown above, that when suitable variations in response to a stimulus appear, these specifically directed reactions should be cultivated to an ever higher degree of perfection by the struggle of parts.

The limit of this increasing lastingness and adaptation on the path to internal recultivation is set only by the potency of the chemical elements of our planet, which perhaps cannot produce the highest degree of perfection for some qualities. However, the reactive substances of our organism are in part already very fine. We need only recall the delicacy of the sense of smell, with which we can perceive less than a millionth of a milligram of musk. We remind the reader of the delicacy of taste, of vision, of hearing, and of the sense of touch that signals to the brain the lightest contact of a downy feather on the foot. This includes a perfection in the reception of stimuli by the end-organs and an unimpeded transmission, which can already be described as fairly perfect.

With the foregoing explanations, the number of cultivated properties that may develop through the struggle of molecules without the participation of a struggle of individuals is not yet exhausted, and we do not even aspire to do this and thereby anticipate the physiologist. What mattered to us here was merely to show the way in which the struggle of molecules works and to demonstrate the "necessity" of cultivating certain properties whose existence we consider necessary to explain certain morphological accomplishments of organisms, which we shall use in the following chapters.

We have thus seen that the struggle of the smallest living parts, as far as it is connected to material exchange, always cultivates the strongest processes under the given circumstances and that under the influence of stimuli it will again select something strengthened by it in some way but without any consideration of specific purposiveness for the whole organism.

[Thus, the struggle, and self-elimination, of parts cultivates the qualities that are most self-sustaining in themselves under existing conditions. The self-preservation of organisms, however, takes place to a large extent through assimilation; thus, all properties that are conducive to this basic organic function are cultivated.]

It follows further that the struggle of parts at the same time ensures homogeneity of cell composition, in that only one quality usually dominates in each cell, unless there are two qualities distinguished by different properties that maintain one another in balance. But because absolute balance will almost never occur and, if it does, cannot exist when circumstances change, the struggle of parts will bring about the greatest possible homogeneity of cellular composition; of course, according to what has been said above, this is unless the cell comes under alternating and different conditions.

This victory of a property, which leads to homogeneity within each cell, has one more essential consequence for selection by the struggle of individuals that should be emphasized. By the struggle of parts, each novel, more powerful quality that appears first as traces gains greater prominence because it spreads out within a particular area: first, in all cells in which it was simultaneously created as a trace, and then, as we shall see further, also into farther areas in which these cells dominate through the struggle of cells. Therefore, if this quality is advantageous for the preservation of the whole, its usefulness will be immediately heightened by the struggle of parts, but if it is disadvantageous for the whole, its harmfulness will be heightened, and thus it will either be more energetically preserved or it will be eliminated more quickly through self-selection.

It is self-evident that no cell substance can possess every victory-conferring property, and it will therefore depend on special circumstances which combinations of favorable properties will gain control. Which of the most varied combinations will lead to victory will be one of the future tasks of physiology to determine.

Here I only want to indicate in what way other properties can exclude all others and persist alone by this principle of the struggle of living molecules (or smallest insubstantiations of life processes).

In addition to the useful properties that triumph in the struggle of parts in material exchange and growth or in the absence of food, other newly arriving properties can even conquer and spread in a direct way: namely, in direct combat with the old, in that the older inhabitants are directly destroyed and consumed by the newcomers [like the surroundings of a carcinoma]; or are assimilated perhaps by a fermentative mode of action; or in a similar way to how a state of excitation spreads in nerves and muscles; or in some other still unknown manner [as I have demonstrated in the direct transdifferentiation of indifferent cells (or already differentiated cells) by adjacent differentiated cells during postgeneration and regeneration].

We observe "pathological" dissemination in the spread [of amyloid degeneration from the periphery in cardiac muscle fibers] in progressive diseases of the nervous or muscular system, such as the progressive atrophy of the spinal cord, in progressive bulbar paralysis, in *paralysis acuta ascendens,* and in progressive muscular atrophy (according to Nikolaus Friedreich and Ludwig Lichtheim), which all spread continuously through connected structures; and also in the way in which the spread of inflammation was previously thought to come about by phlogogenic (inflammatory) action of the products of inflammation; and how Rudolf Virchow has recently considered it possible for infectious diseases to disseminate for which no living agent can be detected; {or how the change in quality spreads within the cells after poisoning with arsenic, phosphorus, or lead or after the introduction of the rabies or syphilis poison into the organism}.[15] Similar kinds of dissemination may have previously occurred, or may still occur, normally in a beneficial way.

It must appear superfluous, given the current limitations of our knowledge, to attempt to make further conjectures about the extent of such processes in physiological events.

Similar processes of the spreading of certain properties through the struggle of parts, like those described here for the cell body or cytoplasm, must naturally also occur in the cell nucleus, only that they are perhaps less, or not at all, under the influence of stimuli.

2. *The Struggle among Cells*

Because, as we have seen, the individual occurrence as such is not firmly standardized and because from the start not all cells of the same tissue will be of perfectly equal life force, a so-called struggle of cells must occur during the period in which the cells of a tissue are still multiplying; for those cells which, under the existing conditions, are best placed to multiply will multiply more rapidly than the others, and thus, with the limited space, more or less take the others' offspring from them, thus inhibiting their further cultivation and proliferation. The more powerful will therefore produce more offspring than the weaker.

---

15 The material in curly braces was present in the first edition but deleted in the second edition. —Trans.

If we ask about the properties that will decide this struggle of cells, we find that they are the same properties already proven in the struggle of living molecules. One can presume that those that regenerate themselves more easily in material exchange due to stronger affinities and likewise those that consume less gain superiority over those less endowed with these properties, *ceteris paribus*. For a better ability to nourish oneself and lesser consumption for one's own needs are certainly to be regarded as favorable preconditions for growth. The same is true of those cells that can best nourish themselves with the quality of available food as well as of those which in the case of greater deficiency increase their affinity for food [that is, those that are capable of self-regulation]. Here, too, if there is a lack of food, those that consume the most food in material exchange are most likely to starve and die out, *ceteris paribus*.

Likewise, among cells that are exposed to stimuli, *ceteris paribus*, those cells will have an essential advantage, and proliferate more, that are *least rapidly consumed by the stimulus* and that are *strengthened by the stimulus in their affinity for food and in regeneration*. Possibly, even more, those cells will overtake the others that are strengthened by the stimulus to the point of *overcompensation*. Then again, cells whose cell substance absorbs the stimulus more easily must gain an advantage therefrom; and if only one stimulus acts on a tissue, the cell quality that is most strengthened by this specific stimulus will be most expanded in the proliferation of cells, *ceteris paribus*.

Because everything happens according to dynamic equivalents, cells will proliferate until each of the cells, over which the total stimulus is distributed, receives only so much of the stimulus that at the median stimulus size it is no longer stimulated to further increase. [A certain organ size corresponds to a constant stimulus size; when the organ size is reached, there is balance between the two and thus stability of form.][16]

Once again, as in the struggle of living molecules, if the quality of the stimulus changes, then a new cell quality will be cultivated from the variations that arise and that can be bequeathed from cell to cell, which interferes with and defeats the old cell quality directly in its nutrition, quite apart

---

16 Roux's argument can be considered a version of the economic principle of equalization of marginal utilities. —Trans.

from the fact that the old cell quality must also fall prey to atrophy by itself because of the disappearance of its own life stimulus. If, on the other
hand, different stimuli act alternatively and repeatedly, then ultimately it
is not a single kind of cell, simultaneously strengthened by the various
stimuli, that is victorious but different kinds of cells produced next to one
another, each of which is strengthened solely by one stimulus but is particularly strongly strengthened by this one. There is a tendency toward
greater and greater specific differentiation in the struggle of cells, as in the
struggle of molecules. For here, too, it is impossible for one quality to be so
intensely strengthened by two different stimuli that it can prevail over two
different qualities, each of which is completely adapted to a single stimulus.
If, therefore, varieties corresponding to the latter relations have appeared,
they must be awarded predominance.

This theoretical derivation of qualities that must eventually be victorious
in battle may at first sight appear exceedingly otiose, but it is not quite so.
First, it is not without use for our knowledge, especially as a heuristic
principle. Second, when we consider actual conditions, there are important clues that the properties we have recognized as winning victory if they
occur, actually do occur. That is, the qualities of the constitutive elements
of our planet have, in fact, been sufficient to bring into being these theoretically derived victors.

Everything appears to proceed the same in the struggle of cells as in the
struggle of living molecules; exactly the same properties are needed for victory and implemented in the same way.

But this would be premature. In the first place, the fact that not only
enlargement but also proliferation of cells must take place causes a difference between the cultivating effects of the struggles of cells and of smallest
living cell parts. For it is possible and probable that different factors increase the mass of cells by proliferation than by enlargement, although the
general preconditions for both are the same, as assumed above.

According to contemporary views, we can distinguish two cellular components: a "cell body" responsible for the function of the cell and its
special life, and another segregated part, the cell nucleus, which initiates
the eventual proliferation [and chiefly implements the formative events].
There are grounds to ascribe quite different qualities to these two parts
because they have different functions.

In any case, those qualities of the nucleus that can best nourish them-selves under the given conditions must also spread more powerfully; and likewise, in the case of stimulus effects penetrating to the nucleus, such qualities of the nucleus gain an advantage through greater expansion that are strengthened by the stimulus in their life process and stimulated to in-crease their substance. All of the above-mentioned "general preserva-tional properties" must therefore also be transferred to substances of the nucleus and must be victorious in the nucleus and victorious with the nucleus.

In relation to these properties of the nucleus, it must remain question-able whether and to what extent the properties of functioning cell bodies can also be regarded as favorable preconditions for the nuclear prolifera-tion that is the starting point of cellular proliferation. Therefore, with our present ignorance of these conditions we cannot judge with certainty how far the struggle of cells cultivates the same things as the struggle of living molecules, how far they promote or resist each other; but we shall find reasons from the empirically observed behavior to conclude that they promote each other. A probability of this can also be deduced from the principle of struggle; for when qualities appear in the nucleus and cell body, which are both strengthened by the same external conditions, the cells thus composed will again have an advantage over others in which only one of the components is strengthened.

[But the cell nucleus, as a result of its ideoplastic functions, as the bearer of the bequeathed formative abilities that typify the self, is better secured than the cell body by morphological self-regulation; so for this reason the struggle among different parts of the nucleus will be very limited as com-pared with the struggle among different parts of the cell body.]

A tendency toward a struggle between cell body and cell nucleus for space appears to remain in the organs of higher organisms, for as soon as the specifically functioning part atrophies in a muscle, a proliferation of cell nuclei takes place (so-called atrophic nuclear proliferation) that does not lead to a proliferation of cells. The same thing also happens, according to Walther Flemming, in cells of atrophic adipose tissue. But these phenomena are of no importance to us at present, and we in no way attach the value of our deductions to the future interpretation of these processes.

We now come to a further difference between the struggles of cells and of living molecules that concerns relations in the "struggle for space." And because we are dealing here with larger relations that can be directly observed, we are also in a position to check more precisely the actual justification of the assumptions.

This is about the relative relations in the "struggle for space." If a contested space is unbounded, then, as described above, faster growth is sufficient by itself for victory, *ceteris paribus*.

Passive cessation of growth is, on the other hand, always brought about by resistance of neighboring parts. So pressure necessarily sets barriers to growth. Although this is very well-known in theory, it has not been proven for animals in the way that is necessary here. The pressure for which this is known is always a pressure that spreads out over a larger area composed of cells and blood vessels; but because the blood vessels, especially the capillaries, are easily compressed, the affected areas of the tissue are deprived of nourishment and are subject to starvation atrophy but not to pressure atrophy.

Of course, such a compression of blood vessels cannot occur in a struggle among individual cells, but the pressure of cells on one another can be mechanically disadvantageous in a similar way; for cells themselves are also traversed by networks of fluid-filled spaces that are narrowed during compression so that nutrition must be impaired.

Apart from its disadvantages, pressure also confers benefits in an internal excitation of diffusion, of protoplasmic movement, or of chemical reactions. And even in the struggle of living molecules the field is not actively vacated by the retreat of vanquished substances, but they must be actively cleared by the more rapidly growing substance or the defeated substances will resist further expansion by counterpressure. Every struggle for space can only be decided mechanically through pressure.

It is now evident that greater resistance to pressure, if it is possessed by a cell and its offspring to a greater extent than by other cells in its vicinity, can likewise lead to victory, even if by a slow but steady general expansion. [Thus, the most pressure-resistant cells are also cultivated in the struggle of cells. Many pathological occurrences show considerable differences in their resistance to pressure; foci of inflammation and tumors that cause their neighbors to shrink feel much harder to the touch.

Rudolf Virchow emphasized the nutritive tautness or tension that is the source of the tone of tissues whereas atony signifies poor external nutrition, with slackening (relaxation) and weakness (debility). The ability of a cell to survive the struggle for space depends on this tone, which, of course, differs among individual cells. But this ability is not the only factor because soft granulations of bone marrow are able to dissolve hard bone substance. Virchow continued, "A growing part, whether it simply swells, or whether it grows rampantly, pushes neighbors apart, deprives them of food, crushes them, and starves them to death. There is a relation of opposition, a nutritive antagonism, between parts of the same tissue as well as between different organs."]

Whether resistance to pressure is already possible within cells, in the struggle of living molecules, will depend on whether the pressure acts purely mechanically or also chemically, and in the former case whether the parts of the newly occurring variation in the cell are firmly congregated together so that as a new component acting as a whole, like a tumor, they can mechanically struggle as a whole against their neighbors. We have no verdict as to the occurrence of this property, and I therefore neglected to mention and use it in my discussion of the struggle of molecules.

We have evidence of the struggle of individual cells against one another in numerous pathological occurrences where cells invade other cells, for example, in lacunar usurations of muscle fibers by sarcoma cells (according to Richard Volkmann and Rudolf Klemensiewicz) or temporary blockage of blood circulation by amoebocytes (according to Richard Erbkam). [Louis-Antoine Ranvier and Guido Tizzoni found white blood cells invading nerve fibers, where they consumed the nerve marrow.] Furthermore, from the normal regeneration of epithelia, as we have come to know for the epidermis (Gustav Lott) and for the ciliated epithelium of the trachea (Otto Drasch), there is most definitely a mutual influence by pressure, which by breaking through, dividing, and dismembering, leads to the partial disappearance and expulsion of the old cells of these tissues.

[These examples demonstrate, first of all, the fact of a struggle of cells. The former cases represent a struggle of cells of different tissues with one another, which can give rise to somewhat stronger properties. The latter examples, on the other hand, concern a struggle of younger cells against older cells of the same tissue; because the victorious younger cells later become older cells, this struggle does not cultivate any permanently stronger qualities. On the other hand, such a cultivating struggle for space perhaps occurs in the vigorous growth of embryos and youths. I believe that I see a manifestation of overwhelming resistance to pressure prevailing in different cells of the same tissue when one cell bulges into another, although it is true that similar forms can be produced by contraction to a sphere, which occurs on the death of cells that have not yet differentiated in shape, such as cleavage cells.]

Elimination of products of material exchange whose accumulation would be harmful is one of the most important general requirements of life—first, when these products take up space for the organism as dead material, especially when these products have a direct harmful effect due to their chemical nature that has become different from the organism. Hence, cells that form less harmful products of material exchange as well as those that can more easily remove these products will have an essential advantage over other cells in the struggle, *ceteris paribus,* and will therefore be more easily preserved and disseminated.

[Differential abilities of cells to migrate can also lead to competition and selection, in that under some special circumstances cells are preserved if they are induced to migrate toward points of higher concentration by the diffusion gradient from a source of food, or if they flee from points of higher concentration of harmful substances. These properties were called positive and negative chemotropism by Theodor Wilhelm Engelmann and Wilhelm Friedrich Pfeffer. Other cells that lack these properties or possess them to a lesser degree do not survive. Thus, positive and negative chemotropism must have been cultivated by the struggle of cells when suitable variations occurred.]

The magnitude of action of the cultivating struggle of cells [(as of the struggle of living molecules) is first due to the appearance of new capabilities of assimilation that are bequeathable from cell to cell and that confer greater lastingness among comrades of the same order given the external factors (food, stimuli, chronic harmful effects) that influence things of this order. Second, the magnitude of action of this cultivating struggle] is determined by the number of cell generations in which it acts, and this, of course, depends on the point in time in the life of the individual at which the newly bequeathable property appears. If it appears only in adults, where only physiological regeneration takes place, it can only work in those organs whose cells are still subject to physiological proliferation either for regeneration or for work hypertrophy—that is, in epithelia, mucous glands (and perhaps other glands), muscles, bones, cartilage, connective tissue, and nerves (but not, as we know so far, in the stock of ganglia or sensory cells).

If, on the other hand, the new variant appears in the early embryo, then the territory it occupies will expand from this fleck beyond that of its originally equal comrades because of its advantage in expansion, and by this means a new favorable molecular variant, even one that appears only after the formation of the blastemas, may almost immediately propagate through an entire [?] tissue. And what is cultivated in this way is what is the strongest for life, what is most suitable for self-preservation, what perhaps most powerfully reacts to stimuli, whether of a physical or chemical nature, with the capacity for work hyperplasia if the stimulus strengthens consumption to the point of overcompensation.

From the general properties thus cultivated, selection in the struggle for existence will cultivate that which is useful to the whole individual (at the same time as the struggle of cells but only secondarily). This cultivation is facilitated by the fact that, as a result of the further spreading of new stronger properties in the struggle of cells, each new character immediately appears with more decisive prominence. When it is useful, it immediately gains more beneficial influence. When it is detrimental, it more rapidly disappears again by self-elimination of the individual from the ranks of the living.

The struggle of parts can also have a direct "formative" effect on cells: on the one hand, it preserves cells that gain a favorable position relative to the blood vessels from which cells ingest nourishment—that is, cells that possess trophotropism, the hereditary capacity to migrate toward sources of food and gather around them—or, on the other hand, and as explained above, cells are preserved that, under the action of certain frequently occurring stimuli, are strengthened in their maintenance and stimulated to growth. These "qualities" will have a shaping effect, provided that these "stimuli" themselves possess a certain limited range; thus, as it were, they have themselves a shape, like the pressure in bones, the tension in tendons, ligaments, and fascia, which will be discussed in more detail in the chapter on stimulus effects.

3. The Struggle among Tissues

A struggle is naturally also possible between different tissues. However, because this is a struggle between heterogeneous things, it cannot lead to selection of the better as occurs in struggles among things of the same kind; it cannot promote or shorten the development of the realm of organisms by enhancing its properties; it cannot "cultivate" what is best [just as little as fights or contests between students and apprentices lead to cultivation of competent officials or skilled craftsmen, for characteristics are often decisive in such struggles of differently acting beings that do not come into play in the specific performances that are beneficial for success in their respective professions].

The enduring result of the struggle of tissues must be a balance among tissues. Tissues that are too vigorous for the well-being of other tissues, even if they themselves would be very useful, must lead to the destruction of the whole, as numerous pathological examples actually show us. Tumors are tissues endowed with abnormal life force that develop at the expense of the food and space of other tissues, and with this destroy the whole organism. The strengthening of connective substances, such as that produced by syphilis toxin, should also be counted here. As is well-known, connective substances increase after the action of this toxin (whether only after the addition of another accidental stimulus is irrelevant here) and finally cause shrinkage of the embedded specifically functioning parts of the organs. In a similar way, arsenic, after long consumption, strengthens the development of fatty tissue as well as other parts. If this increase is often advantageous—for example, in women—its disadvantage is the excessive increase in adipose tissue in general obesity that especially hinders the motion of the heart. Furthermore, every inflammation (in its dissolution and destruction of normal tissue) shows us such a struggle.

One tissue may even win predominance because of the abnormal weakening of another and expanding into the territory of the weakened tissue. Epithelia do this in old age as a result of weakened connective tissue, according to Karl Thiersch. This is how Julius Cohnheim explains the intrusion of blood vessels into old, weakened cartilage, and how atypical epithelial growths penetrate into inflamed connective tissue (according to Carl Friedländer). [Another consequence is that if one tissue is lost, the neighboring tissue becomes hypertrophied at this location, as presented by Simon Samuel and Carl Weigert.]

These examples clearly demonstrate that normal life is tied to a balance of tissues. We see this in other ways; if, for example, an incision is made in the cornea of the eye, the corneal epithelium at first proliferates very quickly and grows into and fills the resulting gap in the connective tissue, but then the regenerating connective tissue gradually pushes out the epithelial bung (according to Hans von Wyss). If a wound is not quickly enough covered by epithelium from the edge of the lost substance, granulation tissue grows out of the open area as so-called wild flesh, but it is held back at the edges, within normal limits, by the delicate newly formed epithelial border.

A lack of balance among the various tissues very quickly leads to death of the individual and thus to the elimination of the unbalanced state and its detrimental quality from the ranks of the living. Only states of balance among the tissues remain in surviving individuals, and thus a harmonious unity of the whole can be cultivated by self-elimination of the deviant.

The balance thus created is acquired only for a certain normal breadth of life and can easily be disturbed by changed conditions. If, for example, connective substances are subjected to an abnormal increase in their blood supply that is not caused by their own activity (such as, for example, occurs in chronic ulcers of the lower leg), then the connective tissue of the skin proliferates, sometimes also the connective tissue of the underlying muscles, and causes the muscles to disappear in the affected places.

It is difficult to judge whether such a balance of forces exists throughout the entirety of normal life or whether in the embryo one tissue must sometimes actively prevail over another to produce normal forms. After a histological study of the lungs, Franz Christian Boll categorically stated the latter as a general principle of all embryonic development, but also stated that he had no idea of the methods that are necessary to establish such a claim. His conclusion that lung and connective tissues each attempt to assert themselves against the other, and that the formation of each organ is the result of a struggle between these two factors, is based on a thoroughly ambiguous study. He used the expression "struggle of tissues" in this sense {and introduced it into the literature}.[17] Because no one will be found to defend this one-sided view of morphogenesis without any evidence, it would be pointless to delve into it further, and so the arrogance of the author in judging the achievements of our most deserving men, which is only matched by his lack of criticism of the execution of his own work, will remain unreproved. As far as he is correct in his view—namely, as far as he thinks that in the embryo parts do not "uniformly" develop "next to each other" but that often in the creation of new forms one part, now the other, plays a more active or a more passive role—this had already been recognized by earlier investigators, and the detailed investigation of these relationships had been tackled [by way of experiment first undertaken by Ludwig Fick and more recently descriptively] by Wilhelm His with his methods for the most precise topography of events.

Because the specifically functioning tissues are always embedded in supporting tissues, which at the same time also contain nourishing vessels and are thereby separated from other specific parts of the same organ, the struggle of specific tissues must always take place first with the connective tissue.

But it seems that connective tissue still has a tendency in many locations to continue to proliferate in adults, for if the specific parts of the glands, nervous system, or muscles perish for any cause, the interstitial connective tissue hypertrophies and occupies the vacated space to a greater or lesser extent [a causal nexus, which was first recognized by Carl Weigert for the connective substance of the kidneys, liver, and muscles, and used to derive the so-called interstitial inflammation].

Occurrences of this kind are particularly evident in cord-like degenerations of the spinal cord, because here the hypertrophy of connective tissue can be regarded with certainty as a secondary phenomenon that followed the atrophy of the nervous parts: first, because the affections were already apparent at the stage of mere atrophy; second, because primary disease of

---

17 The words in curly brackets were present in the first edition but deleted in the second edition. The phrase *Kampf der Gewebe* can be found on pages 35 and 78 of Boll's *Das Princip des Wachsthums* (1876). Boll also wrote, "In the organic world every form is the result of a compromise that ended the struggle between bequeathal and adaptation" (14). —Trans.

the connective tissue affecting individual nerve tracts of the spinal cord or brain would be absolutely incomprehensible given the uniform nature of the connective substance of adjacent nerve tracts. Of course, this does not preclude the possibility that the degenerative process began in the connective substance as the first place of origin (as must be assumed in some cases of spinal phthisis); but in this case the strand-by-strand progression of the process then took place by the destruction of nervous parts that were then replaced by the cord-like compensatory hypertrophy of the supporting substance.

The same thing happens in our bodies with any loss of substance through wounding. In this case, the replacement of what has been lost with connective tissue is usually called "regeneration," but does not deserve this name because it does not restore the normal structure of the site. It is unrelated to the regeneration of lower animals (for example, amphibians) which, according to the views of Carl Hasse and myself, is based on the maintenance of embryonic properties in cells that, after removal of a body part, restores the part normally, except for the species characters.

[The struggle of tissues therefore belongs primarily to the field of pathology. An individual's life is threatened once the normal balance is disturbed. That is why the pathologist tracks imbalances in various diseases and tries to determine how, in each particular case, the perturbation led to death.]

A destructive struggle of adult tissues appears to be a normal process only in bones, in that here the destruction of the old takes place under intrusions of capillary loops. Large multinucleated cells, osteoclasts or myeloplaxes (whose function was recognized by Albert von Kölliker), pave the way for the capillaries by dissolving the bone substance. Something similar takes place in embryonic and juvenile individuals during the destruction of cartilaginous skeletal parts prior to ossification. [To these can be added the physiological pressure atrophy of bones due to the growth of vessels and nerves that penetrate the bone, an atrophy that often reaches abnormal degrees due to veins containing stagnant blood, also the modeling effects of muscles adjacent to bones. Likewise, direct destruction, being eaten up by other cells, by leukocytes, seems to be the norm in the obliteration of older parts of some soft tissues.] {Occasionally a physiological struggle of tissues occurs in other places. According to Wilhelm Krause the *musculus transversus menti* provides an example because this muscle, when it is developed, prevents local accumulation of fat in the subcutaneous connective tissue which then creates the dimple of the chin.}[18]

Despite this mode of action, which is in general modest and which, as we have seen, does not promote the development and strengthening of the organic, the struggle of tissues can in one respect still be of major directly useful importance. In the event that the tissues have the property of strengthening solely from stimuli acting on them, the struggle of tissues immediately becomes a contest between the functional strengths of tissues, and as such endeavors to give each tissue the dimensions necessary for the use that the whole employs of its function. If, for example, the glandular tissue is stimulated to proliferate by a stronger stimulus, then by this increase the supporting connective tissue and the blood vessels are also stimulated to corresponding proliferation. The same is true of all other tissues and their functional interrelationships. If a change in functional conditions leads to tissues being used less often than before, then those tissues will decay because of pressure atrophy from the neighboring parts occupying more space, as well as because of inactivity atrophy.

In this case, the struggle of tissues becomes a principle that directly regulates all quantitative relations within the body, *a principle of functional self-organization of the most purposive proportions.* This principle presupposes that adult tissues receive their vigor by functional stimuli, and develop more strongly and resist their neighbors more strongly with the increase of these stimuli, whereas these properties are weakened by reduced stimulation. Whether and to what extent we are justified in ascribing such important qualities to the tissues of animals will be discussed in greater detail in the next chapter.

4. The Struggle among Organs

In the struggle among organs, heterogeneous parts must once again contend with each other for space and perhaps also for food. Once again, the consequence will be that only such arrangements can persist in which a

---

18 The text in curly brackets was included in the first edition but deleted in the second edition. —Trans.

morphological balance is maintained among the chemically and physiologically quite dissimilar parts. For if the growth force of one organ were so strong that it repressed other organs, then the whole would perish. Perhaps the struggle of organs, like the struggle of tissues, provides the advantage that it rapidly removes the untenable from the ranks of the living, but it must also be considered that it can at the same time also eliminate some of the best-performing substances.

A "shaping" interaction of organs has long been known and appreciated, but not as a "cultivating" struggle for space. The mutual influences of the viscera to shape each other, and especially the passive dependence of the shape of the liver on its neighboring organs, were already observed and emphasized by Vesalius and Jean Cruveilhier (and more recently by Christian Braune, Carl Toldt, Emil Zuckerkandl, Wilhelm His, and others). In the absence of the right kidney and its associated adrenal gland, Theodor Rott found that the pit in the liver that normally accommodated these organs was also absent. The total dependence of the shape of the liver on neighboring organs is even more evident in fishes, in that the liver often insinuates itself among the intestinal loops, filling the gaps like a cast. Similarly, the shape of the lungs depends on the shapes of the chest wall, heart, and dome of the diaphragm; the shape of the adrenal gland depends on the kidney, and the shape of the spleen on the stomach and intestines; and the cerebrum flattens the cerebellar hemispheres. The mutual accommodation of muscles, such as the calf muscles, to each other's shape is morphologically important. If the tongue is congenitally abnormal, the incisors are gradually bent forward by its pressure. When a tooth is extracted, its neighbors gradually approach each other and reduce the gap between them.

Many other reciprocal formative influences could be cited. What is important about them is that this mutual accommodation leads to the greatest possible utilization of space. Therefore, any further enlargement of one organ usually occurs at the expense of another as soon as the latter organ does not possess the power to resist the growth pressure from the former organ and to compel it to enlarge itself only toward the outside [by bulging out the body cover].

If, as already mentioned and assumed in the struggle of tissues, the tissues attain their strengthening of growth and self-preservation solely through the functional stimulus, then the struggle of organs similarly becomes a very useful principle according to the following: first, the organs develop to the size that corresponds to the organism's needs: second, the organs do not slowly succumb to simple inactivity atrophy when use is reduced, but are directly intruded upon by their stronger neighbors and rapidly reduced to a volume that is useful to the organism and appropriate to the new degree of function. At this diminished size, the organ now has sufficient strength to withstand further encroachment by neighboring organs. The latter is exhibited by the *musculus plantaris* of the calf which, according to the diminution of its function in man, has been reduced to a very small structure, dependent in its shape on the other two calf muscles, but which nevertheless in its preserved remnants exhibits a decidedly fresh and performance-capable appearance. It also follows, as a corollary, that organs that are rarely used can be preserved for a long time in places where there is no competition for space, as, for example, the human ear muscles.

[A particularly important but costly adaptation to the available space caused by these processes is the feathering inherent in almost all muscles. By expanding into the space until counterpressure inhibits further growth, the muscles have grown into forms in which the muscle fibers are highly oblique to the tendon, which is why 20 to 30 percent of their power is often lost for their intended accomplishment.

This last struggle of organs for space between adjacent organs of the same tissue type—for example, between two muscles—cannot have a cultivating effect like the struggle for space among cells or similar cell parts because each organ has its own particular activity and corresponding capacity for growth.]

I believe that some of the phenomena summarized by Darwin under the principle of economy of growth can be explained more exactly by this direct struggle of organs for space than if their diminution to a size appropriate to circumstances was selected from accidental variations, as Darwin assumes has been the main factor in their origin.

Organs may fight for food as well as for space. This relationship appears to have long been recognized and correctly understood. Johann Wolfgang Goethe and Étienne Geoffroy St. Hilaire simultaneously proposed laws of compensatory growth: when much organic matter is used to augment a part, food must be withdrawn from other parts, which are thereby reduced. Darwin ascribes little importance to this law relative to the action of natural selection, in that he says, "As the supply of organised matter is not unlimited, the principle of compensation sometimes comes into action; so that, when one part is significantly developed, adjoining parts or functions are completely reduced; but this principle is probably of much less importance than the more general one of the economy of growth."

Such an effect need not take place merely by collateral withdrawal of blood, in the sense that, *ceteris paribus,* neighboring organs receive less blood when the vessels of one organ dilate; rather, such interplay seems to take place in a completely different way. It is well known that women who breastfeed for years suffer from so-called osteomalacia, or bone-softening, which consists of the mammary glands' continuous withdrawal of calcium salts from the mother's bones in anticipation of their replenishment in her diet so that the bone substance softens as it is constantly re-formed. At last, when the old bone substance has been completely replaced by lime-free substance, bones are reduced to a waxy softness and can bend into any shape. Here, then, a struggle occurs between mammary glands and bones in that cells of the former attract calcium salts from the transudate more strongly than does the basic bone substance, thus preempting deposition of calcium in bone.

Overview of the Achievements of the Struggle of Parts

In the preceding chapter we first pointed out that the development of organisms generally takes place according to valid fixed laws but nevertheless does not deliver constancy because these laws are already altered by external influences in the pre-embryo and during embryonic development and even more so in postembryonic development so that a variability in all parts must be assumed, which is confirmed in actually observed behavior. We also stated that as a result of this indeterminacy, bequeathal cannot determine the last individual event but must be satisfied with setting general, normative provisions for what occurs overall.

A necessary consequence of this liberty and dissimilarity among parts was that in material exchange and growth the stronger parts must interfere with weaker parts by depriving them of space, and possibly also of food, and must expand at their expense.

We then saw that struggles of "similar" parts, be they living molecules or cells, "cultivate" [from the variations occurring among them] a series of most-lasting qualities which, because of their general character, are also most useful to an individual in its struggle for existence: first because the most general properties that confer victory in battle are the same everywhere; and second because the whole, as the resultant of its parts, has the same needs as its parts {and can only fight through them}.[19] But as the parts in their battle, one against another, develop themselves to ever greater performance, their overall effectiveness must increase accordingly, in the same way as the performance of an army improves when its officers contend with one another and only the best are selected to train the next generation.

[So, in general, the struggle of parts is based on the same precepts as the struggle of individuals: first, on the material exchange of organisms, which necessitates replacement and thus food; second, on growth, which requires space. Therefore, in both these contests there is a so-called struggle for food (including all the factors influencing nutrition and the maintenance of material exchange) as well as for space.

To maintain material exchange, if we consider each property separately, what is most beneficial is the smallest possible variation and the best possible replacement. All properties aimed in this direction will be most lasting. All properties that deviate from this aim will more easily perish with unfavorable changes in subsequent conditions and thus fade from the ranks of the living by self-elimination. To replace, it is necessary to eat. Only those living substances that are well-maintained with the quality of available food can last under these conditions; the others dwindle away through self-eradication. The same happens with the inclusion of harmful ingredients in the diet: everything that is nonlasting disappears; what remains is the lasting and is therefore resistant, immune.

---

19 This clause {*und blos durch sie kämpfen kann*} was included in the first edition but was deleted in the second. —Trans.

If there is insufficient food to maintain the various organic substances, a struggle for food must take place. This struggle is indirect in that the living substance that is endowed with a stronger affinity for food appropriates more for itself than the living substance with the weaker affinity, which is disadvantaged by the other's existence without the two interacting directly. Moreover, substances that require more food for their maintenance starve to death earlier and disappear through self-eradication.

In addition to consumption and replacement of what has been consumed (pure material exchange) there is also overcompensation of what has been consumed (growth). The latter, as far as it is considered here, consists of a proliferation of living mass and therefore takes place with a spatial expansion of structure. If the available space is absolutely limited or if there are external obstacles to expansion, a struggle for space must occur between different things that expand in the same space. This direct struggle can only be decided by pressure on one side or on the other.

A substance that grows more rapidly has the advantage over a more slowly growing substance that it is dispersed more widely, and is therefore more lasting in the event of local harmful effects, and furthermore maintains a greater share of the activity of life than the other substance.

Thus, the more lasting is always selected but by different processes: food deprivation causes "indirect" struggle for food (material exchange as such leads to the persistence of the more lasting by self-elimination or self-eradication of the less lasting); growth leads to a "direct" struggle for space.

In summary we can say, the variation of similar living parts within organisms leads to a selection of parts in which (by a struggle of parts or by self-elimination) the not-lasting or less-lasting parts are excised and the most lasting are cultivated. Thereby, and in general, the lastingness (so-called purposiveness) of the individual is also raised.]

On the other hand, from the large number of these highest-achieving and most self-preservational properties, the selection of what is suitable for the individual's particular activities in its relations with the outer world must naturally take place solely through the struggle for existence among individuals.

The existing properties of individuals represent only special cases and combinations of those properties that are able to maintain themselves in the struggle of parts. Those substances that prevail in the struggle of parts, but which are unsuited to the preservation of individuals in their relations with the outer world, are eliminated from the ranks of the living, together with the individuals that bear them, and have always thus been eliminated.

Let us also assume that, among the variations in living substances, there were some whose ability to assimilate was increased by the input of a stimulus and on which the stimulus thus exerted a trophic nutrition-promoting effect, either direct or indirect. Such qualities must triumph. The cultivating struggles among living molecules and among cells must then have continually raised abilities to react to stimuli and possibly also have developed abilities to overcompensate for what was consumed by the stimulus, which in turn led to a work hypertrophy; conversely and finally, the stimulus must have become an indispensable life stimulus whose continued presence prevented inactivity atrophy. These two qualities, by themselves, are then able to regulate all quantitative relations in the organism according to the measure of need.

Moreover, given the trophic stimulus effect from the struggle of parts, a principle of differentiation that increases with the stimulus also arose because only those connections can be most strengthened by a stimulus that are directed to it alone and not, at the same time, strengthened by other stimuli; and therefore, properties perfectly adapted to a repetitive stimulus, once they had appeared first in traces, must ultimately dominate.

Furthermore, we concluded that with greater dispersion, which newly emerging stronger qualities acquire through the struggles of molecules and cells, on the one hand, the homogeneous composition of cells and tissues is ensured, but on the other hand, what is more important, the newly occurring variation with the greater dispersion immediately achieves greater significance so that perhaps its benefit becomes more important and decisive in the struggle of individuals; or, in the opposite case, if the property is detrimental, the individuals that are so burdened are immediately excluded from the ranks of the living.

The accomplishments of struggles of "different-kinded" tissues and organs are different. These struggles lead, through self-elimination, to the sole survival of organ qualities that are able to maintain their morphological balance in the body, and further lead to the greatest possible use of space. It can also be assumed that the strengthening of tissues by stimuli enables self-regulation of the quantitative development of tissues and organs according to the needs of the whole.

Accordingly, the functionally necessary proportions develop by themselves through each of the four levels of struggle (of molecules, cells, tissues, and organs): for enlargement by strengthening the ability to feed; and for diminution by weakening the same as well as, in the struggle for space, by the direct interference of parts that are more heavily used.

[Of course, the prerequisite for the "cultivating" (thus "lasting") action of selection is that remaining qualities must persist and be bequeathable so that they can be transferred to the offspring of the selected structures (cell parts, cells, or tissues). When, on the other hand, selection chooses only temporarily more-lasting structures, perhaps the parts of a tissue or cell that are younger or just rested, that have not overexerted, then selection can have no cultivating action, and therefore no lasting adaptation or immunity can result. The magnitude and variety of the cultivating action of the selection of parts therefore depends on the magnitude and variety of bequeathable qualities. The eventual magnitude of action will of course be greater, and therefore easier to determine in the event of major changes in living conditions.]

Finally, it was briefly indicated that the reactive properties thus acquired in the struggle of parts are also capable of "direct functional self-organization" of highly purposive relations of forms; and we promised to set forth the reasons for assuming the existence of such inestimably valuable properties. This will be the subject of the next chapter.

If, as has been done up till now, one derived all good properties of organisms solely from direct selection in the struggle for existence among individuals, then this would be the same as deriving all good institutions of the state and all performance [*Leistungsfähigkeit*] of the representatives of these institutions from the struggle with belligerent neighbors (not only those institutions directly belonging to the military capacity, but also government, legislation, administration, science, art, trade, and industry). With this parable I believed I had already briefly but clearly indicated the significance of the struggle of parts.[20] For who would want to deny that competition and contest among the representatives of the same class, and also the regulative interplay of the different classes on one another, belong to the most powerful factors of steady progress? How far could we have come, just by fighting with neighboring states, without this contest of individuals?

---

20 Wilhelm Roux, "Ueber die Bedeutung der Ablenkung des Arterienstammes bei der Astabgabe," *Jenaische Zeitschrift für Naturwissenschaften* 13 (1879): 336. In this earlier parable, Roux described the cell-states as striking against each other [*der Zellenstaaten unter einander einschlagen*]. —Trans.

CHAPTER III

Evidence of the Trophic Effect
of Functional Stimuli

A. Behavior Due to "Functional" Stimuli and
"Trophically" Stimulated Substances

Of the properties cited in the preceding chapter that must triumph in the struggles of living molecules and cells, or that persist after the self-eradication of other properties, no one will dispute the existence of those properties that simply triumph in material exchange, those that nourish themselves best and consume the least of the available food. First, the premise, material exchange, is an indubitable fact, by which the victory of the favored properties becomes a necessity. Second, the high-grade performance that an organism exhibits directly proves that such excellent properties are present. (Physiology shows us these properties and presents us with the higher organism as the machine that best exploits the tension applied to it.) But it has surely been sufficiently shown that, if these properties are present, they must have spread by the struggle of parts; and by the struggle of individuals only those special cases that are most favorable to the external conditions of the organic species could be selected. These properties are purely physiological and have no particular formative effect, so we have no occasion to go into them further. Nevertheless, a few things will be mentioned in chapter IV about their kinds of occurrence in organisms, as well as about the factors that give the organs concerned their form.

It is different, on the other hand, with the assumption that living substances, or more accurately processes, are present in organisms whose assimilation is strengthened by applied stimuli and therefore must reign in the relevant parts of organisms, provided that these stimuli act repeatedly throughout life. I believe that I sufficiently demonstrated in the previous chapter that such properties would be victorious if they arose. It remains, therefore, to provide evidence that such properties actually occur in organisms before one can finally proceed to an aphoristic presentation of their specific formative performances in the development of the animal kingdom.

In consequence of the difficulty of proving the existence of such qualified materials, it will be best if, in order to be able to recognize their presence, we first derive the general mode of action of these substances and compare them with actually existing conditions.

Processes or insubstantiations of processes whose assimilation is strengthened by increased consumption under the influence of stimuli (in which the original general aptitude for growth, the ability to overcompensate, has remained despite the dependence on stimuli) will be insubstantiated or unfurl to a greater volume with more of the stimulus. Thus, there will be a quantitative self-regulation of the size of organs according to the magnitude of the applied stimulus. As is well-known, all parts of organisms, other than their internal and external surfaces, are protected from foreign stimuli. The stimuli that act upon the parts are therefore just the functional stimuli, such as the neural impulse for ganglia, muscle, and some gland cells; such as pressure and tension for the connective (supporting) substances, bones, cartilage, connective tissues, and elastic tissues.

Thus, when adaptation to an exclusively acting stimulus has occurred through the selection of parts, the more frequently the stimulus acts, the greater the development of each organ. But because these stimuli occur solely as a result of the activities of the whole organism, in that they all depend directly or indirectly on the stimulus center in the brain, they will produce only what is purposive for the whole organism—that is, they will form directly what is purposive for the preservation of the individual. As is well-known, this is actually the case in the organs so mentioned, according to Lamarck, Darwin, and others, and corresponding to my ex-position in the first chapter.

With regard to the magnitude of such overcompensation, we recall the investigations of Alfred Volkmann, who showed that blood vessels can withstand ten to fourteen times their normal tension. With regard to mus-cles, everyone knows that once he began practicing with ten-pound dumb-bells in his youth—which at first he was only able to lift in a particular way with the most strenuous effort of will—after a while he could do this with ease and with the same strong exertion of will could now move fourteen- or sixteen-pound dumbbells with his somewhat thickened arm. We also know from everyday experience that the bones and ligaments are normally able to withstand much greater loads than those to which they are adapted through habitual use. Thus in "overcompensation" we find the first con-cordance of what is factual with what is to be achieved by the hypothetical property.

If the adaptation to the stimulus is so perfect that it has become an indispensable life stimulus, that without it the assimilation and the main-tenance of normal quality do not take place at all, then something further will arise. The living parts will only be able to preserve and develop them-selves "where" the stimulus acts; and furthermore, where the stimulus assumes a specific form, the organs must assume the form and structure of the stimulus, their stimulated form.

If the stimulus acts preferentially in certain directions, as occurs in bones, then mother cells lying in these directions will be most stimulated to formation of bone substance; and because they overcompensate, so much bone substance will be formed in these directions that it will com-pletely absorb the stimulus while for the parts aligned in other directions, if they were formed at all, there will no longer be adequate remaining stimulus. As a result of withdrawal of the stimulus, these parts cannot re-generate and will, as a consequence, sooner or later be lost permanently. In this way, each trabecula of a bone unburdens its immediate surround-ings. And if the directions that are most strongly used are sustained by increased substance, then, as a result of overcompensation, they will also be able to withstand and relieve the tension in other directions that are seldom or less frequently used.

The same thing will take place in connective tissue, indeed in all organs and tissues that have a purely mechanical function and whose stimulus therefore has a definite internal and external form, as we have learned from graphic statics.

If fasciae were composed of completely disordered fibers, those cells and fibers that lay in the direction of greatest pull would be most [intermittently] stretched and strengthened; and because the cells adjacent to these fibers consequently secrete most of the intercellular substance, these fibers will gradually draw off more and more of the stimulus from other cells, preventing them from regenerating and secreting fibers such that finally the latter vanish and only fibers that run in the direction of strongest performance remain. If the tissue, from the outset, has the property of being formed solely under the action of the stimulus, then such incorrectly positioned fibers will only be found weakly developed in the earliest stages of development.

As shown earlier, this is exactly the case for both bone and connective tissue. Both possess structures that correspond to lines of pressure and tension. Whether the stimulus also spreads in a specific way in muscular, glandular, sensory, and ganglion cells and can therefore produce a specific structure of these parts cannot at present be decided, and therefore we cannot at present relate the existing structures of these parts to their functional stimuli.

It was shown in the previous chapter that, if parts adapt to stimuli, different adaptations must occur to different stimuli. And because an organic structure changes a stimulus as soon as it strikes the organic structure—as in the case of the sense organs, where the stimulus is not consumed but passed on in changed form as a so-called excitation—an ever more extensive differentiation occurs through adaptation to ever more finely differentiated stimulating qualities. So it is not merely that a particular quality of the percipient elements must be cultivated, when suitable qualitative variations occur, for each more frequently recurring sensory stimulus, but rather that a whole series of stimuli possibly arise when passing through a series of gradually transforming cells if the first receiving cell does not already possess the ability to transform it into a suitable form for the cells of consciousness. This behavior is presented to us, actually put before our eyes, in the multiple subdivisions into which the transformation of the movement of light is partitioned so that, in addition to the sensory cell, three ganglion cells must be passed through in the retina before the stimulus has attained a quality suitable for transmission and processing in the brain. [The nature and manner of this formation is discussed in the first chapter.]

In the presence of the supposed property and the occurrence of suitable variations, it is further required that particular adaptations to the stimulus intensities, insofar as they recur regularly, must be cultivated, for it is precisely to these stimulus strengths that substances must react most strongly. As is well-known, this is also to a marked degree the case for the stimulated organs (muscles, glands, nerves, sense organs): they all react most completely in their specific manner only to certain medium stimulus strengths, but relatively more weakly to substantially greater or lesser intensities; the same behavior is also evident in the muscles in a particular formal relation, which I intend to discuss in detail elsewhere.

The correspondence of these possible performances of substances, which are trophically influenced by the stimuli, with actual relations in organisms—especially the correspondence of the structures of bones and fasciae with the directions of greatest tension (which as I showed in chapter I cannot be explained by Darwin)—demonstrates very clearly, in my opinion, the qualities we have given these substances.

However, I cannot consider what has been said for passively functioning bones and fasciae as sufficient evidence for other tissues (muscles, glands, nerves, ganglion cells, sensory cells). However, for the organs formed from these other tissues, there are other grounds for recognizing the functional stimulus as having an influence that strengthens their assimilation. For these actively functioning organs, there exists a large series of very interesting and highly significant observations, of which we shall first present those that reveal the consequences of the withdrawal of stimuli after the nerves supplying the stimulus have been severed.

According to the consistent observations of numerous investigators, within a few weeks of cutting a motor nerve, the associated muscle atrophies into a cord of connective tissue with absolute certainty. Ludimar Hermann says in relation to this, "A stable preserving influence of the nervous system is shown by this fact, however much is still lacking for its understanding." After three or four days the direct and indirect excitability of the muscle decreases. The atrophy occurs with indistinct transverse stripes, granular opacity, depletion of the specific substance, accumulation of fat granules, and finally complete depletion of the specific structures. Thus, under the suppression of the normal specific material exchange, another material exchange takes place, of which it is unknown whether it merely represents a halt in the normal material exchange at a lower level or whether it has its own particular quality that directly hinders normal regeneration. According to Moritz Schiff, this process is considerably delayed by regular electrical stimulation of the atrophied organ. On the other hand, it can occur (to a lesser extent even without cutting the nerves) simply by completely abandoning use of the muscle, as is a common side effect in surgical diseases—for example, chronic inflammation of the joints or large tumors. It seems to me to follow from this that the functional stimulus is indispensable for the maintenance of muscles; and Julius Cohnheim also says, "The elements of the working organs assimilate only when they are stimulated, not by mere hyperemia."

The fundamental fact that glands are acted upon by nerves was discovered in 1852 by Carl Ludwig for the submaxillary gland, and then extended by other authors to other salivary glands, and more recently by Balthasar Luchsinger to sweat glands. Experiments with nerve cutting have shown that not only can stimuli trigger glandular function, but the functional stimulus is also necessary for maintaining the normal condition of the gland. After excising a segment of the testicular nerve, Auguste Nélaton (and after him Ivan Nikolaevich Obolensky) found fatty degeneration of the distal segment of the nerve and tubular epithelia of the testis and epididymis, with gradual complete shrinkage of the epithelia (despite preservation of the vessels) and spatially compensatory formation of fatty connective tissue. Similarly, Friedrich Bidder and Rudolf Heidenhain observed a very rapid diminution in size and reduction in consistency of the submaxillary gland after cutting through its nerves. Twenty days after the cut, Bidder obtained glands weighing 8.7 grams on the cut side and 15.5 grams on the side with uncut nerves. [This example was inapt because atrophy arose as a result of persistent paralytic secretion, which occurs after a few days, and thus as a result of exhaustion.] Luchsinger found that, six days after cutting through the sciatic nerve, pilocarpine (which otherwise acts directly on the

cells of the sweat glands) is no longer able to induce perspiration, probably as a result of the degeneration of the gland cells which occurred after
cutting the nerve. [This statement has been retracted by its author.] That
these consequences of nerve cutting in muscles and glands are not due to
alteration of the blood supply will be explained further below.

If sensory nerves are cut, investigators believe that, in the same way as
after cutting motor nerves, it is the peripheral segment of the nerve that is
separated from the center, which rapidly atrophies, whereas the central segment remains intact for a long time. However, when sensory nerves are
cut, their end-organs, the sense organs, remain permanently intact (unlike
the degeneration of muscles when motor nerves are cut). Wilhelm Krause
has repeatedly established the latter for the optic nerve, and Paul Langerhans has most recently. Langerhans has shown it also for the tactile cor-
puscles of Langerhans.[1] Georg Meissner and Krause, however, disagree
with Langerhans because they observed atrophy in these organs after cutting the nerves. Giuseppe Colasanti also observed degeneration of olfactory cells after cutting the olfactory nerve. However, these conditions are
very difficult to observe, and the consequences of such interventions are
perhaps more complex than we can presently imagine; therefore, we must
reserve our judgment. Nevertheless, the preservation of sense organs after
cutting their nerves would speak in favor of our view that the specific
stimuli are at the same time the preservers of life activity because the specific stimuli are still acting upon the sense organs after the nerves have been
cut. Experimental interruption of sensory stimuli does not appear to be
possible for most senses. Only for the eye could one determine whether
the retina would atrophy after suturing the eyelids, then pulling over and
suturing the skin from the neighboring parts and keeping the animal in the
dark. This experiment has not yet been performed; but perhaps a great
experiment of nature can be interpreted in this way. The eyes of animals
that live in dark caves have degenerated or completely disappeared. [Path-
ological cases also indicate the preservation of retinal responsiveness to
light through the action of light: eyelid cramps are associated with temporary blindness.]

---

1   Now understood to be antigen-presenting dendritic cells of the epidermis. —Trans.

According to Augustus Volney Waller, the atrophy of the distal segment of a nerve after it has been cut occurs very quickly and is very complete, in that the axillary cylinder and the nerve marrow disappear within a few days; whereas, as mentioned, the proximal stump remains intact. Experiments by Alfred Vulpian, Moritz Schiff, and others with double cutting of a nerve again show that only the part still connected to the central organ persists, so degeneration and loss must be regarded as the result of the separation from the central organ.

Further experiments by Waller with transection of the posterior, sensitive spinal root showed that afterward the entire peripheral nerve was preserved and the central stump deteriorated, from which can be inferred that the sustaining force for the sensory nerves does not come from ganglion cells of the spinal cord but from those of the intervertebral ganglia. These experiments definitely demonstrate the sustaining influence exerted on nerves by ganglion cells.

For motor nerves, which do not enter into any connection with ganglion cells in the intervertebral ganglia, on the other hand, according to the findings and analogy with the sensory nerves, the sustaining force must come from the large ganglion cells of the anterior horns of the spinal cord because these ganglion cells are the only ones that stay in contact and communicate directly with the nerves. This view is further confirmed by the pathological occurrences listed below in which these ganglion cells were destroyed. The central stumps of the nerve remain intact for years after transection, apart from the fact that, according to Wilhelm Engelmann, on the central stump the content of the fiber always curdles right up to the next node of Ranvier and dies. According to Wilhelm Kühne, this preservation should take place here and under normal conditions by direct feeding of the nerves from the ganglion cell. But in my opinion it is absolutely impossible that a meter-long thread of microscopic fineness, which is often very active, is fed from one end, especially because this thread itself consists of hundreds of individual threads, the primitive fibrils of the axis cylinder, from which arises an almost insurmountable resistance to the movement of material parts over long distances.

I therefore consider the hypothesis of other authors to be more probable, that neural nourishment takes place by penetration of food at the nodes of Ranvier, and I assume that ganglion cells merely provide a sustaining life stimulus.

It should also be mentioned that nerve swellings and neuromas often occur at the nerve endings in amputation stumps. According to what has been said, we shall be able to assume that these have arisen from the damming up of the life stimulus emanating from the ganglion cells that will result in increased nutrition; and perhaps a greater or lesser proportion is to be ascribed to direct excitations from mechanical insults from which they are struck from the amputated surface.

It should not go unmentioned that the preservation of the central stumps of sensory nerves, in spite of their apparent intactness, is not perfect, for they gradually lose their excitability; thus, the stimulus emanating from the ganglion cells alone is not sufficient for maintenance, but it appears at the same time that the specific functional stimulus also has a preserving effect.

It can also be assumed that the stumps of central motor nerves are still receiving weak functional, emanated stimuli because stimuli are very easily dispersed in the network of ganglion cells in the spinal cord. One often sees that, when learning movements that are difficult to perform, one frequently moves muscles that cannot contribute anything to the intended movement. Just as, in these cases, stimuli disperse into the wrong pathways, weak stimuli will occasionally penetrate the peripherally interrupted pathways if neighboring ganglion cells are innervated. This happens perhaps more frequently and more generally than we presently suppose because we merely pay attention to impulses that are strong enough to induce contractions. The following statement by Ludimar Hermann will perhaps find confirmation; he says, "The muscle may have degrees of excitation that are expressed in chemical or galvanic processes, but not yet in contraction processes; and with the nerves it is even probable that they have excitation processes that are insufficient to induce muscle contraction."

According to François Magendie, after transection of the optic nerve not only the peripheral but also the central segment degenerates. Hermann remarks that perhaps the cause is the fact that this nerve has no nodes of Ranvier.

If transected nerves heal (which always occurs by sprouting from the central stump) then the peripheral stump, which has meanwhile undergone fatty degeneration, quickly returns to normal, as the fat granules disappear and become normally excitable and conductive. The specific processes are thus probably strengthened by the accustomed stimulus so that the prior changes again become insubstantial and disappear. [According to recent observations, the peripheral segment of the nerve is not revived, but a new shoot sprouts from the center to the periphery, using the earlier nerve only as a tube for guidance.]

In addition to these important experimental results, some relevant pathological occurrences should also be mentioned, which likewise show the consequences of withholding functional stimuli.

So-called spinal childhood paralysis is characterized by severe atrophy of the muscles connected to the affected ganglion cells or nerves. This disease of the nervous system consists mainly of the destruction of the motor ganglion cells of the spinal cord, but occasionally also begins with a disease of the peripheral nerves that deprives the muscles of their functional stimulus. It follows from this that the functional stimulus is necessary for normal development, even for young, still-growing muscles. Therefore, development does not take place purely through bequeathed properties of these parts. Furthermore, there is an equivalent disease in adults that is also associated with atrophic paralysis according to Hermann Eichhorst and Ernst Leyden. In general, associated muscles atrophy when the spinal cord is affected.

We are right to conclude from these experimental and pathological observations of degeneration and loss of muscles and glands that have been deprived of the functional stimulus that the functional stimulus in these organs causes not only dissimilation (using up of materials) but is also essential for the assimilation required for regeneration. The functional stimulus must have a similar effect on the nerves themselves, but not sufficient in itself for maintaining their material exchange. As we have seen, however, the greater part of their preservation must be due to a stimulus emanating from the ganglion cells.

As mentioned above, such an effect of the stimuli has already, in recent times, been assumed by Ewald Hering and used to explain phenomena of sight and sensation of warmth; only he allows assimilation and dissimilation in these organs to be stimulated by distinct stimuli.

Trophic effects of stimuli that promote the nutrition of the parts and are supposed to be necessary for normal nutrition have long been accepted by physiologists. It has been thought that these stimuli are conveyed to the parts by special trophic nerves. Indeed, some authors are inclined to ascribe a widespread occurrence and correspondingly great importance to trophic nerves. However, although I accept the trophic effect of stimuli, I must object to the existence of separate pathways for these stimuli, and in this objection I fully agree with Sigmund Mayer, who has recently discussed the question in detail.

Trophic nerves were first inferred from the effect of cutting the *trigeminus* (sensory nerve of the face). Subsequent inflammations of the eye and ulcers in the oral cavity, which occur frequently, were ascribed to disturbances of the nutrition of these parts. It has, however, been established by the investigations of many researchers, such as by Alexander Rollett, and more recently by Hugo Senftleben and Nathaniel Feuer, that these inflammations are a consequence of the loss of sensitivity and thus the failure to remove harmful substances through protective reflex movements.

In addition to these inflammatory changes, other trophic disorders occur after nerves are cut. Moritz Schiff found the bones of a leg were thinner and the periosteum thicker when its nerves (*nervus ischiadicus, nervus cruralis*) were cut. Alfred Vulpian and Max Kassowitz found something similar. According to the above, the thinning of the compact bones is more likely to be traced back to a lack of functional stimulus as a result of the paralysis of the muscles and an inactivity atrophy caused by this paralysis (or according to Christian Nasse, to the narrowing of the blood vessels) than to the action of particular trophic nerves, for which we have no anatomical understanding and for which we do not know physiologically where this formative stimulus could be produced and how it could produce the correct formations. Self-organization through the effect of the functional stimulus, on the other hand, appears to be the simplest and most obvious explanation. The thickening of the periosteum with irregular formation of bone can probably be attributed to the expansion of blood vessels after transection of the nerve, for there is occasion to ascribe the ability to grow more with increased blood supply to the bones and connective substances.

Hermann Joseph's and Hermann Schulz's findings of similar changes in both hind legs of a frog after cutting the nerves of one hind leg and plastering both hind legs to a perfect position of rest certainly speaks in favor of our view. It also explains the result of Moritz Schiff that the leg bones did not become thinner after cutting through the *plexus ischiadicus* (nerve plexus of the leg) of a frog that he galvanized daily for six months; this caused the muscles to contract daily, thus saving them from atrophy and at the same time preserving the bones under almost normal functional conditions—that is, under the action of muscular tension. [The contraction forces are more important for sideways leg positions of this animal than is the force of the weight of its trunk because the contraction forces are only transferred to the bone via the action of muscles.] After cutting the nerves of a hind leg of a pregnant bitch, Schiff found bone softening (osteomalacia) as well as bone thinning after the suckling period only on the side of the transected nerve. This is readily understandable. If normal formation of bone substance is linked to action of the functional stimulus on bone-forming cells, then calcium deficiency must affect the paralyzed leg first.

Furthermore, the atrophy of submaxillary glands and testes that occurs after transection of glandular nerves in no way necessitates the assumption of special trophic nerves through which special nutritional stimuli are supplied, for it is certainly easier to assume that the functional stimulus has a trophic effect that strengthens assimilation. The same is true of the muscular atrophy that occurs after cutting a muscle's nerves or after pathological deterioration of the nerves or associated ganglion cells of the spinal cord.

Just as the foregoing experiments can all be explained as inactivity atrophy resulting from a lack of functional stimuli, without assuming particular trophic nerves, one could also attribute the consequences of cutting sympathetic nerves (of the visceral and vascular nervous systems) purely to vascular disturbances: *in part,* enhancement of the blood supply, which is sufficient for increased nutrition if not for increased function, in the supporting substances (connective tissue and bones) of young and adult individuals and also in the working organs, especially the muscles and glands of young individuals; *in part,* atrophy as a result of anemia.

It is a matter for physiology to discover why, after a sympathetic nerve has been cut, sometimes narrowing and sometimes widening of blood vessels is observed—thus, sometimes atrophy and sometimes hypertrophy. Atrophy was observed by Moritz Schiff after cutting the sympathetic nerves supplying the wattles on a turkey's throat. After extirpation of the upper cervical ganglion of a young cock, Charles Legros observed atrophy on the corresponding side of the crest. After transection of a cervical sympathetic nerve in guinea pigs, Charles-Édouard Brown-Séquard found that the brain had become clearly atrophic on the corresponding side (confirmed by Alfred Vulpian in one case). On the other hand, Friedrich Bidder, Moritz Schiff, Sigmund Mayer, and others obtained heightened growth after cutting sympathetic nerves.

Perhaps the nutritional disorders observed in neuralgia (nerve pain) and other pathological cases can also be attributed to vasomotor disorders. Neuralgia is accompanied by changes in the number, color, thickness, and spread of the hair, thinning of the skin, shrinkage of the fat pad, and also skin rashes (herpes, urticaria, pemphigus, etc.). The same disorders also occasionally occur with peripheral anesthesia (numbness) as a result of peripheral interruption of neural conduction. As a result of peripheral paralysis, the skin often becomes atrophic (paper-thin, smooth, and shiny on the fingers and toes) and prone to decubitus[2] and ulceration. Hypertrophy of skin and nails and increased growth of hair has also been observed: Silas Weir Mitchell observed loss of hair; Paul Schiefferdecker observed greater growth of hair. Bone atrophy and liver affections also occur.

Acute decubitus (Jean-Martin Charcot), which often occurs in injuries to the spinal cord and which quickly spreads even with the greatest possible protection from pressure and with the greatest possible cleanliness, seemed to speak in favor of trophic nerves. But because here above all the skin and the underlying connective tissue die, on which no one has ever seen nerves approach the cells or fibers, the direct action of the nerves, apart from the vascular nerves, seems to be the least well-founded [?]; and it is probably

---

2 *Decubitus,* bed sores, pressure ulcers. —Trans.

more correct to look around for all conceivable alternative causes for these cases, as well as for the *hemiatrophia facialis progressiva* (unilateral facial shrinkage), than to assume a new nervous system with a source of stimulus, of which there is no sign, and its stimulus regulation.

Also to be mentioned are diseases of the joints in peripheral paralysis, in injury to the spinal cord in *tabes dorsualis* (spinal cord dizziness), in spontaneous inflammation of the spinal cord, and in hemiplegia due to brain affection. With our present imperfect knowledge, all of these can be traced back more or less to inevitable consequences of paralysis and do not necessitate the assumption of particular trophic nerves. But small hemorrhages (found by Moritz Schiff, Charles-Édouard Brown-Séquard, and Wilhelm Ebstein in the lungs, stomach, and pleura after brain injuries to the striatal nuclei, optic nuclei, and pons) indicate peculiar vasomotor disturbances from changes in the central nervous system.

In summary, I believe that changes observed in muscles and glands result from the loss of functional stimuli due to neural affections, but those in passively active organs (bone and connective tissue) must be attributed mainly to vascular disturbances. For the latter, however, it should not be forgotten that they are also subject to inactivity atrophy, especially the bones.

This does not, however, exclude the possibility that some organs receive trophic stimuli conducted by particular nerves, but these would then not be general but special, limited arrangements. Hermann Eichhorst, for example, assumes that trophic stimuli, which are indispensable [?] for maintaining heart muscle, are supplied to the heart via the path of the vagus nerve. Moreover, we found it necessary above to allow the ganglion cells of intervertebral ganglia to exert an indispensable, sustaining influence on sensory nerves, even if, as we have seen, the ganglion cells alone (without the stimulus of the specific function) are unable to preserve the excitability of the nerves.

Rudolf Heidenhain deduced from the peculiar behavior of the submaxillary gland after poisoning and stimulation of the *nervus lingualis,* that the path of this nerve contained special secretory fibers, which he called trophic fibers, in addition to the vasodilatory fibers. These secretory fibers, which stimulate turnover of the organic constituents of the glandular cells, perhaps act not only on dissimilation (more rapid glandular secretion) but may also act directly or indirectly to increase assimilation, because otherwise exhaustion must occur immediately after cellular reserves are depleted. They would thus be trophic nerves in the sense that we postulate: they would be trophic and functional at the same time, and the stimulus they convey, be it physiological or artificial, would possess this double action. The existence of other nerves in the gland that influence the secretion of water does not alter this interpretation.

The functional nerves of other glands and muscles are also to be understood as trophic nerves in our sense. On the other hand, the functional stimuli for connective substances are mechanical and do not require neural transmission. Some other glands, and probably also the kidneys and liver, are excited by chemical stimuli in the blood and therefore do not require functional or trophic nerves. Sensory stimuli could be regarded as playing the same role for sensory cells.

In general, Sigmund Mayer[3] is of the same opinion with regard to the value of the hypothesis of particular trophic nerves and also ascribes a certain trophic influence to the functional stimulus for muscles and glands. He says regarding the glands and muscles, "The central nerve substance (gray matter), the peripheral fiber and its peripheral end-organs represent not only a functional excitation unit, but also an alimentary or nutritive unit."

He also writes, "This hypothesis can explain why nutritional disorders develop in nerves and muscles when the normal connection between them is broken. After such a separation, each part succumbs to its own fate, to put it this way, while the purposes of the organism had closely linked its fate to that of other apparatuses. With the dissolution of the excitation unit, the nutritive unit also disappears. The processes that then develop are not immediately atrophy, but rather allotrophy. The nutritional processes in nerves, muscles, and glands that are disconnected from their centers do not cease but are directed in paths that are no longer subject to the purposes of the whole organism as in a functional relationship. Such a disconnected muscle is paralyzed only for normal movements serving the

---

3 The extensive quotations are from Sigmund Mayer, "Specielle Nervenphysiologie," in "Handbuch der Physiologie des Nervensystems," special volume, *Handbuch der Physiologie,* Ludimar Hermann, ed. (Leipzig: F. C. W. Vogel, 1879), 209–212. —Trans.

purposes of the organism, but otherwise moves spontaneously (paralysis tremors) and remains excitable for artificial stimuli (electricity), albeit in a different way."

He then expresses the opinion that "the central nerve substance does not unilaterally determine the nutritive destiny of the peripheral structures" but rather "is also influenced in its own nutrition by the peripheral organs with which it forms a unit of excitation."

But he admits "that the peripheral apparatuses suffer more easily when the central nerve substance is altered than the other way around. This seems easy to explain when we consider that the peripheral nerve, muscle, or gland is a member of only a single excitation unit. As soon as this unit is destroyed, the normal nourishment that depends on the integrity of this unit must also suffer. As is evident from many observations, the central substance is manifestly a member of miscellaneous functional and nutritive units. If, for example, the connection of a motor nerve with the spinal cord is cut, we see the peripheral stump of the nerve, together with the muscle, falling prey to allotrophy. The central stump and spinal cord remain intact for a long time, probably for no other reason than because the deficit of nutritional impulses in the spinal cord caused by the loss of the muscle and a segment of its nerve can be overcompensated by the intimate connection of the relevant part of the spinal cord with other parts of the peripheral central nervous system and body."

We agree with Mayer apart from his assumption of an invigorating retrograde influence from the muscular and glandular cells to the nerves belonging to them, which does not seem to be established. [His conception, however, lacks the most important of our morphological deductions that the functional stimulus is the formative factor that exerts the trophic effect, be it direct or indirect, so that atrophy must occur not only when the nerve is cut but also when the nerve is externally intact but fails to function.] Therefore, for the reasons given above, in place of the less specific term "nutritive and excitation unit" that is morphologically not equally applicable, I put the precise hypothesis: *In addition to stimulating its specific function, the functional stimulus "directly or indirectly" stimulates assimilation which cannot properly proceed without its influence and thus simultaneously acts trophically to increase nutrition.*

Of course, other processes of material exchange take place in the absence of this stimulus, of which, however, it is unknown whether they are of a peculiar nature and whether they can gain control over the old normal processes, or whether they actively affect them in the struggle of molecules for space and for food, or merely represent a pause at a lower than normal level of material exchange, or whatever is their nature. However, here we are beginning to draw conclusions beyond what we are capable of doing in this publication.

Except in the doctrine of trophic nerves, the trophic effect of stimuli has been assumed since ancient times in theories of the origin of tumors. A discussion of tumors does not really belong here because we would be dealing with abnormal formations and abnormal stimuli. However, we do not want to omit a cursory look at the cause of their origin, in order to possibly obtain a useful analogy to the trophic effect of the surrounding functional stimuli. Only then will we move on to the last, apagogical part of the argument.

Since ancient times (and it is still a very widespread opinion at the present time), surgeons and physicians have assumed that tumors arise from a single stimulus or, more readily, from repeated exposure to stimuli over a period of years, and then, even after the stimulus has ceased, the tumors continue to grow indefinitely by themselves until they have destroyed the organism. Thus, it has been assumed that mechanical or other stimuli can have a very significant trophic effect.

Against this view, Julius Cohnheim has recently rightly emphasized that a stimulus that acts only on the blood vessels will cause increased blood supply to the affected part and hypertrophy (a simple enlargement or increase of the parts) but cannot cause unlimited growth. We agree that the unlimited growth of tumors involves a steadily progressive growth and proliferation, not merely an expansion of the blood vessels as the stimulus can well produce; and it is impossible to see why the growth of tumors, even if it could be provoked by stimuli, which we cannot see at all, should continue without cessation after the stimulus has ceased. [The cells must therefore regain embryonic properties by the stimulation, and thus acquire the ability to continue to grow for a long time without any further stimulation, simply as a result of the "triggering" of immanent forces of growth caused by the stimulus, be it that this property is only formed by the stimulus or, perhaps more conceivable, that embryonic active constituents are activated. But there are insufficient analogies to make this assumption, even if some schools of pathology tend to make it.]

The same applies if cells of parenchyma, rather than blood vessels, are stimulated directly by the functional stimulus. Here, too, the cause remains incomprehensible. Why is there a progressive overpersistent effect? How can overcompensation in assimilation, induced by a stimulus, continue after the stimulus has ceased? Because it has never been possible to establish such a thing with certainty by observation, and because the percentage of those tumors for which stimuli are presumed to have been the cause is only 14 percent, I cannot grant legitimacy to the entirety of Cohnheim's doctrine. However, I agree with him when he extends the idea (previously expressed by Rudolf Virchow and Albert Lücke for special cases) to a general principle that all tumors characterized by unlimited growth are to be regarded as surplus remnants of embryonic tissue which only later express their preserved embryonic peculiarity of progressive growth when the surrounding tissues are sufficiently weakened that they can no longer offer enough resistance. Thus, these tumors cannot provide any analogical support for our conception of the trophic effect of functional stimuli.

But these tumors of unlimited growth are different from another, special group of tumors, namely Virchow's tumors of infections or granulation tumors, to which belong the neoplasms of syphilis, leprosy, tuberculosis, typhoid, and lupus. Here I cannot agree with Cohnheim's view that these tumors—which initially appear in various parts of the body as small nodules made up of round cells in the connective tissue after a demonstrable poisoning of the body with a specific toxin—are caused solely by local dilation of the blood vessels, because we cannot imagine that the effect of circumscribed dilation of the blood vessels would be such that so many cells develop in one place that they gradually compress their feeding vessels as a result of which these neoplasms must perish, as occurs with typhoid and tuberculosis.

Because I furthermore assume, contrary to Cohnheim, that chemical and mechanical stimuli are capable of causing an increase in the cells of connective substances, in the same way as happens in plants through the sting or venom of a gall-wasp or through colonization by aphids, so too we believe that the specific poison acted here as a stimulus to increase. The particular localization and nodular shape of the tumor are just as difficult to understand if one assumes capillary hyperemia to be the cause; for in the latter case it cannot be seen, with the chemical nature of several of these toxins, why there is only a limited capillary hyperemia and not a more extensive hyperemia. The continuation of the process up to the compression of the blood capillaries remains understandable in both cases; for even when the cells multiply *in situ,* growth can continue so long as the capillary remains minimally open and can still give off nourishment, if only the parts themselves are sufficiently stimulated to take up nourishment. These tumors also do not have the character of unlimited growth; and their further formation, as well as the further persistence of what has been formed, seems to be abolished after eradication or removal of the causative toxin. Thus, it seems to me most probable that they arise in the same way through the stimulating effect of a specific toxin, as we are certainly familiar with from goiter. This arises when predisposed individuals enter a region of the goiter, and further growth ceases; indeed the tumor itself sometimes disappears again after they leave the region.

We should then have to recognize examples of the trophic effect of stimuli in the so-called infectious tumors. Indeed, they are probably chemical not physical stimuli, which does not exclude the possibility that in some of these diseases they are produced by microorganisms as, not without a certain justification, is suspected.

B. Insufficient Formative Effect of "Functional Hyperemia"

We shall proceed to the last part of our demonstration of the trophic effect of functional stimuli, to the apagogical proof;[4]—that is, to the exclusion of the effect of "functional hyperemia," the anemia that occurs with the failure of function, which has hitherto been regarded as the cause of functional adaptation by most authors. We shall show that these alterations in the supply of blood cannot explain the phenomena of functional adap-

---

4 An apagogical proof is a *reductio ad absurdum* that shows the rejected alternative to be laughable. In this section, Roux ridicules the theory of "functional hyperemia" by arguing that it requires that the cells of the capillary walls possess almost magical powers. —Trans.

tation, and therefore cannot eliminate the necessity of the principle of trophic stimulation.

It has been asserted, or tacitly assumed, that an increase in blood supply during function and a short time thereafter is the cause of the enlargement of the organ that develops with persistent reinforcement by function. [In 1844, well before Herbert Spencer, Julius Vogel wrote in his article "Hypertrophy," in Rudolph Wagner's *Concise Dictionary of Physiology*, "It has long been recognized that continued congestion is the essential causal factor of hypertrophy." He expresses the cause of activity hypertrophy himself: "The result is probably that the increased activity of the muscle fibers through reflection causes an expansion of the capillary vessels, which is anyway a condition of congestion, which results in increased separation and nutrition."]

It seems obvious that an increase in an organ's ability to take up nutrients requires an increase in nutrition channeled toward it. And since Carl Ludwig and Ivan Sczelkov have actually demonstrated an increase in the blood supply accompanying the function of the most active organs, the muscles—that is, a "functional hyperemia"—it was but a small step to conclude, given the evidence of this connection, to equate the necessary precondition of increased uptake with the *causa efficiens* and to claim that hyperemia is the cause of the increased uptake of nutrition, of the increase in size of the organ—that is to say, of hypertrophy. [Ludimar Hermann says of the increase in volume of heavily used muscles, "A sufficient explanation is missing; the closest is the hypothesis that the periodic hyperemia associated with contractions, as well as the mechanical effects of stretching, form the middle members."

As will be shown below, we cannot agree with this deduction.[5]]

On the one hand, the necessary precondition and cause of a process can in principle be very different from each other; second, an increase in the

---

5 [Herbert Spencer, for whom George Romanes, as stated above, claimed the priority of my deductions, also derives functional adaptations from functional hyperemia. By demonstrating on the following twenty-five pages that this derivation is incorrect and then developing a new theory that does justice to all the facts on the basis of two other principles, I believe that I have, for my part, contributed enough that Mr. Romanes would have had reason to impart this to the readers of his review, instead of leading them, by concealing this fact, to the conclusion that my work is merely a detailed elaboration of some of Spencer's ideas.]

blood supply during or after function has not been demonstrated for all organs; finally, an increased blood supply for increased activity, both in principle and in fact, is not absolutely necessary. It is only necessary not when the usual surplus of nourishment is not present but when the organs' routine nourishment from the blood vessels has been fully utilized. The existence of an excess of nourishment is possible only if nourishment does not depend solely on the influx of food but also on other factors. Conversely, an increased influx of food is absolutely necessary for increased nutrition only if the blood supply is at the same time the source of nutrition. In this case, always as much is taken in as there is, but this case first must be demonstrated. With the usual conclusion that increased blood supply is absolutely necessary for increased nutrition, what is already assumed to be established is what is first to be demonstrated, namely, that the necessary precondition is at the same time a *causa efficiens.* If this is not the case, there may be an excess of unutilized food. Only in one very special case would the effect of these different causal relationships coincide: the minimum amount of food is always supplied such that the greatest possible utilization of the opportunity for nourishment would always take place, even if the nourishment was obtained from other sources. But experience teaches us that we usually have an excess of blood, so we can tolerate considerable loss of blood; thus, the organs are normally supplied with a superabundance of blood.

After this discussion of the principle, let us now turn to actual behavior.

From the fact that for supporting tissues (bone, cartilage, and connective tissue), a functional increase in the supply of food has not been proven, it does not yet follow that it does not occur. We must, therefore, leave this question undecided, and then we cannot directly oppose the hypothesis that increased activity is always connected with increased blood supply.

When two phenomena are always observed together, it is very difficult to recognize in what relation they stand to one another, which depends on the other, or whether both jointly depend on a third factor; for logic only teaches us that phenomena that always occur together must be causally connected in some manner. But we are no longer in this uncomfortable position because we now have observations at our disposal that show that these two phenomena can be separated.

First, we know that hyperemia does not induce function, neither in muscles and nerves in which function is indispensably linked to material consumption, nor in glands, those organs in which the products of material exchange perform the organismic function. Although for glands material exchange is itself the essence and the purpose of their function, thus the possibility that the supply of substances directly triggers the function seems to be particularly obvious, Paul Keuchel demonstrated that, after poisoning with atropine, stimulation of the *nervus lingualis* does not cause an increase in secretion despite hyperemia of the submaxillary gland. Finally, increased blood supply cannot induce the functioning of supporting organs that function only passively.

Second, and conversely, it would be possible that the function would cause hyperemia, an increase in the blood supply. This possibility seems to correspond to the actual circumstances in many cases, and it will therefore be discussed in more detail in the following. The relationship, however, is not absolutely fixed such that the function cannot take place without causing hyperemia; for after poisoning with physostigmine the blood vessels are not dilated when the lingual nerve is irritated, but the secretion is increased; and Balthasar Luchsinger found that pilocarpine can induce perspiration on the hind paw even if the abdominal aorta is blocked and the circulation is thereby blocked. Naturally, however, this function cannot last longer than the exhaustion of the glands because the possibility of regeneration is eliminated by cessation of the circulation.

The third possibility was that function and hyperemia are not in a direct relation of dependence on one another, but that both jointly depend on a third relation. This type of connection seems to occur in muscles and glands. At least some authors assume that with the impulse of activity for these organs, an impulse for the expansion of the blood vessels immediately emanates from the central organs.

It is sufficient here to have mentioned and distinguished these possibilities. Furthermore, it is better not to go into this until further context has been discussed because for us it is not so much a question of the type of causal connection between function and hyperemia than of the cause of the stronger nutrition that occurs with stronger function. This increased nourishment can only depend on greater food supply, provided that parts always take in as much food as they are offered or, if this is not the case, on a greater intake with the same food supply—that is, on stronger forces of attraction and assimilation. Above all, a decision must now be made between these two options.

Observations on whole persons show that when you put more food into a body, it gains weight, to a particular extent for each individual. This is the case both in adult human beings and to an even greater extent, *ceteris paribus,* with the same degree of function, during the period of independent growth—that is, during youth. When a juvenile or full-grown organism exercises a certain activity with ample food, it gains more weight than with the same activity and little food. Thus, the food intake of the parts of the body is *ceteris paribus* dependent on the amount of food offered.

On the other hand, we observe that this has its limits. One can barely accelerate a young person's growth by copious food. In the same way, with provision of higher quality food, greater development of adult organs in specific parts occurs only within certain limits beyond which it does not proceed, apart from the accumulation of fat. Correspondingly, Julius Cohnheim says that increased food supply does not lead to increased protein storage in the blood or in the tissues if more work is not done at the same time. As is well-known, Carl von Voit found that combustion increases (recognizable by a greater excretion of urea) with a greater supply of protein to the body, *ceteris paribus,* and that only a relatively small increment is retained in the body, and this is for the most part circulating protein as a supply of food, not extra protein in organs with increased protoplasm of cells.

Just as the whole body can spurn ingestion, the substantial assimilation of necessary food, so too can the individual parts of the body.

Many years ago, Rudolf Virchow emphasized the importance of experiments that cut the jugular sympathetic nerve. After such operations, Virchow, Moritz Schiff, and others have observed widening of the blood vessels for weeks without any thickening of the skin or increased peeling. After the same operation, Claude Bernard, Louis Ollier in fifteen cases, and Julius Cohnheim himself observed no hypertrophy even in juveniles. It should be mentioned, however, that only in rare cases does the resulting hyperemia of the skin of the head last a long time, but usually disappears after a few days or weeks.

Similarly, I saw in a doctor and writer an expansion of the blood vessels of the skin on the little finger pads of both hands so that they looked dark pink, without any thickening of the dermis or epidermis and without increased flaking of the latter, although this condition had lasted for seven years. Since attention has been drawn to it, such behavior of the tissues in the case of chronic dilation of the vessels as a result of affections of the vascular nerves (vasomotor neuroses) has often been observed in recent times.

Experiments in which hypertrophy arose cannot prove anything against this ability of the parts to spurn food; they merely show that in other cases, the essential differences of which are not known to us, hyperemia can produce increased uptake. Thus, Alfred Bidder and William Stirling obtained more considerable growth of the ear on the operated side after the above experiment, and subsequently Moritz Schiff and Sigmund Mayer observed that the hairs of the ear grew more rapidly on the operated side, with simultaneous cutting of the auricular and magnus nerve.

James Paget transplanted the spur of a rooster onto the rooster's crest and saw it growing in an extraordinarily strong manner on this vessel-rich tissue. But because the cockscomb itself does not grow continuously in spite of this excess blood, this also suggests that it does not grow merely in accordance with the amount of food offered but that something else is necessary for its ingestion of food.

It is well-known that organs grow more strongly with a greater food supply in the period of self-sufficient inherited growth, also in youth (although, as mentioned, not proportionally and only to a certain extent). Particular causes must therefore be sought for the deviating results in some experiments.

But there also appear to be some tissues that, even when fully grown, can be stimulated to grow again with artificially induced hyperemia—that is, by increasing their food supply. Various pathological phenomena testify to this. Perhaps the thickening of connective tissue, which we find surrounding and deeply beneath chronic ulcers of the lower leg, penetrating deep into the muscles, can be traced back to such long-lasting hyperemia. Similarly, in chronic hyperemia of the skin, hypertrophy of the skin is observed in both the connective tissue and epithelial layers, and increased bone formation is observed in hyperemia of the periosteum.

However, in these and similar cases of inflammatory hyperemia we do not know whether chemical or mechanical stimuli are not also effective in stimulating the increase. But, to use this uncertainty to our disadvantage, we will assume in the following that the supporting substances (bones, cartilage, connective tissue) as well as covering epithelia (those without secretory function) are able to multiply by increasing their food supply without further stimuli. The same has been said for the lymph glands, spleen, and kidney. Because, however, there is reason to assume that the stimulus for the specific function of these organs lies in the blood, with an increased blood supply they are thus stimulated to increased functioning; and the resulting hypertrophy can therefore be regarded as an activity hypertrophy.

The behavior of the passively functioning supporting organs is therefore to be separated in principle from that of the actively working organs (muscles, glands, nerves, ganglion cells, sensory cells) which are not stimulated to hypertrophy or hyperplasia by increased blood supply alone.

The hypothesis could therefore be advanced, at least for the passively acting parts, that functional hypertrophy in them is due to functional hyperemia. But, as mentioned, it is precisely for these organs, with the exception of the skin, that functional hyperemia has not been demonstrated; and besides, as shown above, their structure cannot be deduced from the blood supply because it is partly much finer than the mesh size of the capillary network and shows no resemblance to the arrangement of the capillaries.

Let us now see to what extent the hypothesis of passive cellular nutrition is justifiable and what contradicts it. Even in the fertilized egg, after the formation of the blastemas but before the formation of blood vessels, where food is still evenly distributed, typical uneven growth takes place, which leads to the formation of the primitive streak, the formation of the medullary tube, the axillary cord (*chorda dorsalis*), and the proto-kidneys. Therefore, the ingestion of food must be unequal because the parts experience the same nutritional conditions but grow unevenly to produce specific forms. And because they develop qualitatively differently, a qualitative and quantitative choice of food must take place. This inequality in the attraction of food must be all the greater because the various parts of the germinal disc are not all, as was assumed above for the sake of simplicity, located precisely on the same level as regards the source of the food, but rather the most rapidly differentiating and growing parts of the germinal disc, which are situated next to the axis, are farthest away from the source of food.

The same thing is manifested in bloodless lower animals such as our indigenous small aquatic polyp, the hydra. In these animals, as is well-known, specific morphological differentiations take place through growth of unequal strength in the formation of the tentacles, although the uneven growth cannot be due to unequal distribution but only unequal consumption of food.

On the other hand, however, to ascribe uneven growth in the embryo after the formation of the blood vessels to differential distribution of food by the vessels would mean that the laws of growth reside only in the blood vessels. It would mean that the specific parts do not unfold independently according to inherent laws resulting from a specific nature [thus not by self-differentiation of the "morphological" external influences] but are shaped and enlarged merely according to the distribution of food. The actual laws of growth would then lie in the blood vessels, and the specific cells, which must select specific food from the general nutrient fluid, would be completely dependent on the amount of intake, on the supply alone.

But because the blood vessels, which distribute the food, themselves consist of cells, which must grow and ingest food, so as far as the larger vessels nourished from the *vasa vasorum* (blood vessels that feed blood vessels) are concerned, the laws of growth of the organs reside in the *vasa vasorum;* but insofar as the laws of growth concern the *vasa vasorum* itself and the other small vessels of the body, which nourish themselves directly from the blood flowing within them, these laws must reside in the cells that form the capillary wall; for these would have to take in more food and grow more strongly before they could widen the vessel or create new capillaries. So, in the last instance, greater "active" uptake of food by certain cells must again determine development in the embryo and growing individual.

Following Hermann Fischer, most authors have found that in congenital asymmetric gigantic stature the arteries were not detectably wider in the limb on the larger side of the body than in the normal limb on the other side of the body, and that long-term compression of the arteries on the enlarged side was unable to decrease the size of the limb. It is therefore an entirely unwarranted proceeding to attempt to derive the morphological differentiation of the organism, the development of all the countless individual forms, from the unequal distribution of the blood, even if the latter may occasionally be a favorable factor.

On the basis of pathological observations, Virchow already presented a similar view in his *Cellular Pathology* where he writes, "We are therefore always compelled in the end to regard the individual elements as the operative factors in these attractions. A liver cell will attract certain substances from the blood which flows through the nearest capillary vessel, but the liver cell must first be present and then be able to exercise its very special peculiarity in order to exert this attraction."

To this I shall only add that it is irrelevant for our purposes whether a cell is able to attract specific substances directly from the capillary, or whether the cell only absorbs these substances from the surrounding lymph and, as a result of this removal from the lymph, these substances more rapidly diffuse from the capillary than other substances that have not been removed, and whether the capillaries of the various organs have finally adapted to this stronger penetration of preferentially used substances so that the diffusion resistance for such substances has also been reduced in these organs.

If active growth of the cells through greater "active" intake of food is an indisputable prerequisite for all differentiations, it is certainly more obvious and simpler to relocate this differential activity, and thus the laws of growth and development, to those parts that have the specific qualities—that is, to the specifically functioning cells of the organs and not to the indifferent cells of the capillary wall, which are more equal in all organs or are only secondarily differentiated. We must trace the typical differentiation of form in organisms back to the independent quantitative and qualitative nutritional choices of cells [for the most part] of the specific cells of each organ. Robert Remak acted very carefully when he kept his eye on the differentiation of the specific parts of the organs and regarded them as primarily formative, contrary to the above mentioned, generally quite unmotivated assertion of Franz Boll.

If embryonic and postembryonic development is determined by active quantitative and qualitative food selection by cells, should a completely different, almost opposite law suddenly apply to activity hypertrophy in the still young and in adult individuals? Is nutrition supposed to suddenly become purely passive in the adult, dependent solely on the regulation of the blood vessels, which is now only dependent on nerve mediation from some center?

With the functional enlargement of organs, there is not only a simple enlargement of the elementary parts but also a proliferation in the elementary parts and a proliferation of capillaries. Here again, if the nutrition was purely passive, the capillaries must suddenly be driven to grow more strongly and sprout; and because, according to the law of dimensional activity hypertrophy established in the first chapter, the organs must only enlarge in those dimensions that strengthen their function, the capillaries also must develop only in these directions; the laws of specific formation would once again have to reside in the cells of the capillary wall because a mere increase in the blood supply to the organ with passive growth of the capillaries and the specific parts assumed to be dependent on them would result in an even increase in all three dimensions. But how are the capillaries to be stimulated merely to multiply in two dimensions, excluding the third, through the increased function of an organ?

If the growth of organs is only slightly determined by the blood vessels, but, conversely, it is predominantly the specific parts of the organs that determine the consumption of food by active selection, then the question arises how, under these circumstances, the regulation of an organ's blood vessels always attains the width appropriate to the needs.

So as not to anticipate a special presentation in another place, I do not want to discuss in detail the difficult morphological problem that has preoccupied me for many years of the regulation of blood vessels in the embryo. But it still must be remarked that I do not consider the width of vessels to be bequeathed for only a very few vessels, but that I believe that almost all of them must be understood as determined and developed by means of self-regulation by the consumption of the parenchyma.

A few demonstrative examples will be given here to ground such dependence of blood vessels on the self-sufficient, actively nourishing specific parts of the organs.

Even if the development of the vessels within a tumor could be regarded as congenital in the *potentia* of tumor germs, this would be less likely for the development of the blood vessels that lie outside of tumors that supply and drain them. And should these latter vessels always grow first and only then provide the opportunity for further enlargement of the parts located in the tumor such that the tumor remains in absolute dependence on them?

The objection that the blood vessels develop according to bequeathed formative laws is certainly not possible for the development of the vascular network that develops around parasites after their immigration. If such a thing, for example, an *Echinococcus,* becomes lodged in any organ, it evidently attracts nourishing fluid from a molecular distance, thus causing a constant outflow from the blood vessels with a gradual increase of the capillaries and thus compelling the host, with whom it resides, to surround it with a capillary network and associated larger vessels and thus provide necessary nourishment to its mortal enemy. It is inconceivable that the accumulation of fluid in the *Echinococcus,* and especially its growth, simply takes place mechanically by diffusion, as in dead substances, because for this to be possible the inserted microscopic embryo would have to contain abundant salts, indeed many times greater than itself; but even then the salts would soon all be gone, and a standstill would be established.

Someone might object that the blood vessels of the envelope around the *Echinococcus* that are provided by the host are usually not large. But we see the same thing more clearly in the development of metastatic tumors. If a few cells of a malignant tumor, which have gotten into the blood vessels and been carried away with the blood, become lodged somewhere, they feed themselves there and force their surroundings to form nourishing capillaries and furthermore to form larger blood vessels for further growth of the tumor. Here, too, we have a morphological self-regulation of the blood vessels depending on the rate of consumption of the tumor, both of the vessels within the tumor itself and the larger vessels of the normal surroundings that supply and drain the tumor. The tendency to form these vessels cannot be bequeathed because metastases can lodge at any location.

[If, in this case, the dissemination of tumor germs occurs within the bloodstream, we also have an example of equally successful dissemination outside of the bloodstream. For example, if a cancer of the liver or stomach has reached the surface of the organ, parts of the tumor mass soon detach and are distributed in the abdominal cavity. After these detached parts have lodged somewhere, blood vessels sprout into them, perforating the peritoneal epithelium and feeding them. These "disseminated" tumors often develop to a size many times greater than the original adherent tumor germ. Blood vessels are likewise formed after transplantation of living masses, even dead masses, into already formed or artificial body cavities that are not lined with true epithelia. However, dead masses only become permeated by blood vessels after living cells that eat the dead matter have migrated into it and themselves require nourishment.]

The same is shown by the development of the embryo in the womb. Where the embryo settles and draws food from the mother, her capillaries multiply, and blood vessels, supplying and removing blood according to need, are formed through self-regulation, not just in the uterus but also at any point in the abdominal cavity where the embryo implants in extra-uterine pregnancies. In all these cases, a tendency to develop blood vessels at these specific locations cannot be congenital, but there must be a general reactivity of the organism, according to which blood vessels form themselves everywhere corresponding to local "consumption" as a means of self-organization and morphological self-regulation.

I believe these examples prove that parts actively nourish themselves and that the organism reacts passively with the formation of capillaries and the corresponding supplying and draining vessels. It is extremely difficult to explain how the morphological regulation of the supplying and discharging blood vessels occurs. It in turn presupposes reactive qualities that we have never before suspected. Assuming these few qualities, however, the purposive development of the width of the blood vessels in the whole body is immediately explained, as well as in the pathological neoplasms and parasitic tumors we have mentioned (as which the fruit in the womb must here be regarded as well).

It seems that this regulation of blood vessels, which we see as being entirely dependent on the needs of the parts consuming the blood, is also mediated by newly formed nerves when the vessels become larger; for the smooth muscle fibers, which the vessels of metastatic tumors also have, are probably also supplied by nerves. This neural assistance in regulation is also evident after blood vessels have been ligated. New detours are not only developed mechanically, but at the same time a new neural regulation must also arise, which likewise cannot be bequeathed but has to be developed by means of self-organization and self-regulation.

Just imagine what would have to arise if nutrition were purely passive, entirely dependent on the blood supply, and if, after ligation, the distribution of blood was compensated solely by mechanical collateral effects. What functional disturbances and transformations of the whole part must arise! If, for example, the humeral artery were tied, the shoulder muscles and skin over it would become bulky, and the forearm thin and weak, but none of this occurs; regulation is usually very rapid and perfect. [This applies to all cases in which supply is not provided by so-called end-arteries—that is, vessels with no collateral connections with neighboring arteries.] And because the muscle groups in question later return to full function, a new neural regulation must also have developed for the production of functional hyperemia, which, however, can only arise in direct dependence on the rate of consumption by the parts. All these relations point to self-regulation by need, so we have to assume that the specific parenchyma themselves regulate both the uptake and supply of their need, and that the neural regulatory apparatus also develops in accordance with need and is subordinate to it. For just as they are produced under completely new, unbequeathed conditions in which they can arise merely as a function of the consumption of the parts, they must also be able to develop in a similar manner in normal conditions.

[It is a very widespread but quite erroneous view that the (alleged) congenitally stronger development of the right side of the body is derived from the fact that the arteries on the right arise somewhat closer to the heart than those on the left, and that the former therefore receive more blood; this presupposes that the width of either the right or left vessels is inappropriate. In addition, the latter would not matter at all, even if it were the case, because one finds in some well-formed people as a coincidental finding that the aorta at the point of attachment of the *ligamentum botalli*[6] has shrunk to such an extent that the lumen has completely or almost completely disappeared. What would the body look like if nutrition was passive and the blood distribution was done roughly mechanically according

6   The *ligamentum botalli* is the fibrous remnant of the fetal *ductus arteriosus* which functions to bypass the lungs during fetal development but usually closes at birth. —Trans.

to the size of the vessels? The lower half of the body should resemble a newborn, the upper half a giant.]

We should also like to mention that, as with the other organs, *two kinds of functional regulation or adaptation* must be distinguished in the vessels: first, the merely *transient,* changeable regulation, often mediated by neural function, in the respective *exercise* of the specific function of the organs; second, the *morphological,* long-lasting regulation arising from growth of the organs and their nerves as well as from structural changes. The latter type of adaptation takes place by mechanical mediation, at least initially without nerve mediation, in the embryo and at the beginning of the development of the parasites and tumors that have been mentioned.

The neural regulation of vascular width to comply with the consumption of organs can come about, first, by dividing the functional stimulus between organ and vessel so that a part of the stimulus always passes over to the vessel. As we shall see, however, stimulus qualities are extraordinarily diverse in the body, and the smooth muscle fibers of the vessel wall must always react in an appropriate manner to each of these various stimuli, either by being struck directly by the stimulus or by part of a stimulus being fed to them by special pathways. For organs that are chemically stimulated to activity, nutrition and proliferation must both be stimulated under this stipulation, as in the compensatory hypertrophy of a kidney after excision of the other kidney. The smooth muscles of the renal arteries must react to the accumulation of urea in the blood by relaxation to the same degree as the epithelia of the renal tubules are thereby stimulated to increase their material exchange and function.

However, the prevailing current opinion is that the hyperemia, assumed in this compensatory hypertrophy, is essentially what is known as collateral hyperemia caused mechanically by increased blood flow to neighboring parts after a region of the vascular network has been closed off. Thus, as a result of the occlusion of one renal artery, more blood must flow into the other renal artery for mechanical reasons. But this explanation of hyperemia is incorrect; for no more blood would flow into the other renal artery than its width relative to that of the abdominal aorta and the other vessels originating in the area [and these do not receive more blood than their width relative to the other branches of the whole aorta and the resistance in the whole vascular system of the body against expansion as a result of increased pressure]. Instead of the development of severe hypertrophy only of the kidneys, all organs that draw their blood from this region—that is, the entire lumbar region, the large intestine, the testes—would undergo a little hypertrophy. [One immediately sees that hypertrophy of the organs merely as a result of a mechanical collateral increase in the blood supply would be an entirely useless, indeed an extremely disadvantageous principle.] But in reality none of the consequences corresponding to this principle can be observed.

After removing one testis, there could be no hypertrophy of the other because the occlusion of such a narrow artery and its small capillary area from the entire area supplied by the abdominal aorta, which represents almost half the body, could only increase the blood pressure in the area immeasurably little, and of this increase only the most minimal part would benefit the other testis. Nevertheless, as is well-known, the remaining testis sometimes hypertrophies to a very considerable extent. In addition, the compensatory hypertrophy of lymph glands of the rest of the body, which always occurs after these organs die in one part of the body, could not be explained at all by collateral hyperemia, for how could collateral hyperemia affect very distant small organs in other parts of the body? On the other hand, if we assume that the functional stimulus causes hypertrophy, it arises quite automatically, for that quality of blood which induces the activity of the lymph glands will have a correspondingly stronger effect on the other lymph glands after part of these organs have disappeared.

It should also be mentioned that nerve-mediated regulation of the arteries is so powerful that the occlusion of arteries that are already quite strong can be completely compensated, whereby the effect of mechanical collateral hyperemia on the arteries can be more or less neutralized.

The explanation in terms of collateral hyperemia must be based on the hypothesis that in the various organs that are stimulated to activity by chemical constituents of the blood (kidney, liver, testes (?), spleen, lymph glands) the muscle cells of their blood vessels must react to these chemical stimuli with corresponding strength; while in those glands that are stimulated to act by nerve mediation—for example, the salivary glands—a part of this stimulus should branch off and pass over to the vessels. The same should happen with the muscles and even with the ganglion cells of the brain and spinal cord.

All of this appears extremely complicated. Everywhere the smooth-muscle fibers, which are physiologically identical in all organs, would have to react to special stimuli with a certain appropriate strength; and we would have no understanding whatsoever as to how regulation could arise in novel circumstances.

It is also unthinkable how such a regulation could be active for the bones, because how could the stimulus that strikes the bone also proportionally affect the blood vessels? Or what would be the process in the central nervous system?

If certain nerve tracts or ganglion cells are used more and thus need more food, an individualized neural vascular regulation would have to be there for each fiber, for each ganglion cell, and at the same time it would have to be ensured that the stimuli do not diffuse further because otherwise all neighboring parts would also become hypertrophic. I merely point to Helmholtz's example (discussed above) of rapid functional adaptation, in which, when we see with glasses that transpose two images, thousands or millions of ganglion cells and their branches adapt themselves so quickly that with a few minutes of practice after then removing the glasses we already reach, against our will, in accordance with the way we have just learned. Should this thousand-fold change in nerve connections arise passively through hyperemia, what would be directed into these thousand-fold pathways? Or if it is objected that this is merely the consumption of reserves already stored in the cells (although, as stated, such storage would directly contradict the principle of passive nutrition in which all reserves are exhausted), we only need to think of another example of more sustained exercise, such as playing the piano.

If, on the other hand, hyperemia that is so completely restricted to these individual tracts is impossible with purely passive nourishment of the tissues, then the co-development of the simultaneously hyperemic neighboring parts would make any acquisition of special artistic skills impossible. [As early as 1842, Karl Friedrich Canstatt states in his article on atrophy "that every cell has a peculiar individual life, by virtue of which it, on the one hand, continues to develop itself (plastic power of the cell) and, on the other hand, the substances attracted and absorbed by its cytoblastema[7] are able to change the cell's inner and outer surfaces specifically and chemically (metabolic power of the cell)."]

On these grounds also I am of the opinion that the ingestion of food takes place actively, according to stimulation by the functional stimulus, and that vascular regulation, including mediation by vascular nerves, where it takes place at all (namely, only if larger cell groups take in more food at the same time), will generally be dependent on the consumption by the specific parts of the organs, whether direct or indirect.

As to the manner in which this regulation takes place, I do not want to preempt the physiologist by expressing hypotheses; however, relatively simple modes are conceivable. For the morphological vascular regulation that delivers lasting formations, which alone belongs to my discipline, I hope, according to my present observations, that I will one day be able to trace it back to mechanical principles.

From what has been said about the action of blood distribution and the manner in which it is regulated, it follows that it would have contradicted all that we know if a "passive" diet of the parts had been established, dependent solely on the supply of food; but it turned out that, on the contrary, nutrition takes place with qualitative and quantitative selection on the part of the nourished parts, and that blood supply must become regulated in some way from the "place of consumption" according to its consumption.

"Functional hyperemia," where it occurs, can therefore in no way be the "cause" of functional hypertrophy; but it can only be viewed as a favorable, perhaps not always indispensable, "precondition" for the same.

If we take another look at the possible performance of blood distribution in the absence of function and the "inactivity atrophy" that follows, the causal relationship is apparently simpler here, and the dependence on the blood supply seems to be greater and more definite than in hypertrophy. For if food is supplied in a substantially reduced quantity, dietary intake must of necessity decrease accordingly.

---

7   The fundamental constituent of cells, roughly synonymous with protoplasm. —Trans.

But this is the question: Why does the food supply decrease; why does it not stay in a middling state, since the tension of the blood column here, as everywhere, strives to widen the existing paths instead of making them more narrow? This steady narrowing, which goes beyond what can be purely mediated by nerve regulation, this real morphological regression first needs an explanation that can only be found when, because of the atrophy of the specific parts, the capillary region is reduced by a lack of its function.

But quite apart from this conception of the origin of vascular shrinkage, how should the capillary network of the blood vessels explain the inactivity atrophy of individual nerve tracts in the spinal cord (for which blood can be supplied from all sides) as there is a common, coherent capillary network for the entire cross section of each of the six strands composed of many thousand nerve fibers? In order, by reducing the blood supply, to cause these cord-like atrophies, which are limited to certain nerve tracts along the entire spinal cord, each nerve fiber would have to have its own closed capillary network with independent regulation. The same applies to the atrophy of unburdened trabeculae, which occurs after a broken bone has healed at an angle, when a structure corresponding to the new static conditions is formed.

Just as the fine structural detail that is present in the central nervous system, in bones and fasciae, and in cavity-enclosing muscles could not have developed as an effect of blood distribution through functional hyperemia because the blood distribution in the network of capillaries is not regulated and self-contained in the necessary manner and the mesh size of the capillaries is coarser than the relevant structural detail, so also a thus-restricted food deprivation cannot cause the atrophy of microscopically small, sharply circumscribed parts.

However, to look for special reasons for each organ, for a phenomenon that affects all organs and organ systems, is hard to defend. A specific hypothesis has been attempted for muscles: stretching the sarcolemma (muscle-fiber skin) during contraction favors the passage of food and thus causes activity hypertrophy whereas inactivity atrophy is caused by the absence of this favorable stretching. This hypothesis does not appear to be well-supported, not to mention that it is difficult to provide the same accessory factors for each of the other elementary parts (ganglion cells, nerve fibers, bones, etc.). It is certainly worthwhile to look for and consider such factors, but with such a general phenomenon they can only ever have the significance of accessory factors.

So neither "activity hypertrophy" nor "inactivity hypertrophy," nor the origin of "functional structural detail," can be deduced from regulation of the blood supply in this way. The origin of these relations as a direct or indirect consequence of the "trophic effects of functional stimuli" thereby gains an even greater probability.

The trophic effect of the functional stimulus remains the sole cause of activity hypertrophy, of the overcompensation that enlarges organs, helping to enlarge them with the same structural detail merely according to the dimensions that render them hyperfunctional. For because the parts are not active without the trophic effect of the functional stimulus and degenerate quickly if it is completely excluded but exhibit hypertrophy in its presence, this hypertrophy must now be regarded as a consequence of the strengthening of life processes by the stimulatory effect, because there is no passive nutrition caused by the supply of food.

Finally, the trophic effects of stimuli in general enjoy widespread recognition. Virchow writes:

"We have it in hand to develop both the whole individual and, in particular, some of their organs and systems, and thus to develop their individual peculiarities in this or that direction.

"Among the means of attracting more human flesh, blood, and nerve mass, the most decisive factors are stimuli, the means of excitation. Without stimulus there is no organic work, no absorption of formative substances, no development.

"Salts, spices, certain spirits and volatile substances bring to the organs an excitement that determines them to take up substances, which awakens their internal and external activity.

"Mechanical impulses, the action of light, heat, electricity, and numerous other influences that affect the sensory nerves or the circulating juices or the tissues themselves have the same effect. Above all, it is the spiritual excitement that gives the greatest results (not just thinking, but also being active, willful impulses)."[8]

---

8 Roux quotes from Rudolf Virchow, "Über Erblichkeit" (On inheritance), *Deutsche Jahrbücher für Politik und Literatur* 6 (1863): 339. —Trans.

Résumé

In the chapter on the struggle of parts, it was deduced that living substances in organisms, which upon stimulation experience not only functional change but are simultaneously strengthened in their ability to absorb and assimilate food, must, for general dynamic reasons as soon as they had once appeared in traces, attain dominance and exclude all others in the struggle of parts. Because there was reason to suspect that these properties would be of great significance, not only physiologically but also morphologically, we undertook to verify that such substances, satisfactory in theory, had actually come into being and indeed existed in organisms.

For this task we were compelled to travel separate paths for the two major categories of organismic constituents. First, for supporting substances, especially for formations of bone and connective tissue, we could point out—in the quantitative development of the related organs in their internal structure and external form as well as in their behavior in novel pathological conditions—that an identity exists between the actual performances of these tissues and those that could be theoretically derived. Given the multifacetedness of the performances in which this identity was revealed, an accidental coincidence as a result of other causes could be excluded. Therefore, we were able to conclude from the identity of performances a corresponding identity between the assumed quality and the properties of the supporting substances.

Second, for working organs, for whose structure no possible concordance with the stimulus form can be demonstrated because the form of the stimulus is unknown, we took a divergent path, equally safe, that had been paved by experiments of many excellent researchers. The description of the effect on these organs of keeping the functional stimulus away from them showed us that this causes degeneration, regression, and the disappearance of the specific parts. Therefore, we had to recognize that functional stimulus has a preservational effect and also the effect of strengthening assimilation.

Finally, we discussed the hypothesis, often expressed and at first glance not improbable, that activity hypertrophy and inactivity atrophy are simple consequences of the hyperemia that is present when function is present but absent when function is absent. In consequence of the fundamental importance of this hypothesis and the difficulty of assessing the individual actions of two factors that almost always occur together, activity hypertrophy and its underlying problem of the nutrition of parts were first discussed at greater length. We showed that nutrition cannot be purely passive (simply conditional on food supply) but must depend on the internal conditions of cells in such a way that cells are able to refuse an increased nutritional supply caused by expansion of the blood vessels; to increase intake or keep it constant when the blood vessels constrict; and when there is a constant rate of delivery of food, to sometimes take up more and sometimes less nutrition. We have also seen that the supply of blood to embryonic organs must depend on the state of the organs' specific parts, which must be able to regulate the supply of blood to themselves according to their own consumption. We also saw that this is probably true for the nerve-mediated regulation of blood supply in later embryonic and postembryonic life.

Once we had cut the ground from beneath the feet of this hypothesis, activity hypertrophy could no longer be regarded as an effect of functional hyperemia. Equally, inactivity atrophy could not be regarded as caused by the absence of functional hyperemia. Instead, the former proved to be a consequence of the functional stimulus strengthening the ability to assimilate, and the latter a consequence of the absence of this stimulus weakening assimilation. Functional hyperemia appeared to be no more than a favorable, perhaps dispensable, precondition for functional hypertrophy.

With the validation of the trophic effect of the functional stimulus, "functional adaptation" can thus be traced back to a mechanical principle in its two known groups of achievements, the effects of increased and decreased use, and in the newly established group of "functional structures" of organs. Hence the outstanding effectiveness of organs, which are directly purposive down to the finest structural detail, yielding indeed the most appropriate proportions all the way down to the ultimate living molecules, is no longer to be understood as a teleological but as a mechanical effectiveness.

CHAPTER IV

Differentiating and Formative Effects of Functional Stimuli

It is the spirit that creates the body for itself.

—Schiller

Function makes the organ.

—Guérin

As suggested by its title, this chapter presents conclusions that can be deduced from what has been set out in the two preceding chapters. These conclusions can of course only have a subordinate value because they are based on statements that still require acceptance, and are intended only to show where the principle of the direct or indirect trophic effect of functional stimuli, that I have introduced into morphology, can lead and to stimulate the tackling of solvable, newly arising questions with the means available to us today.

No event can be unilaterally conditioned; every change of state must be brought about by an additional changing force. This applies also to the differentiation of organisms, both in the narrower sense of their shapes as well as the so-called qualitative, or textural, differentiation of tissues.

A. Qualitative Effects of Functional Stimuli

1. We shall first seek to discuss qualitative differentiation, the development of the basic qualities, that is the *origin of tissues.*

Each type of tissue must originally have had a particular first cause. This raises the question whether present-day tissues still require this cause for their development or whether all qualities are now simply transmitted directly by bequeathal and no longer require their original cause.

Bequeathal, the transfer of the chemical qualities of parents to children as parts of themselves, is no longer a problem but is a mechanical necessity.[1] Assimilation causes bequeathal to be a mechanical necessity despite material exchange, for this enables the law of inertia to be transferred from simple physical processes to processes connected with material exchange. Instead of bequeathal, then, the problem becomes rather the development (or forming forth) of what is chemically and morphologically differentiated from what is simpler, merely with a supply of nutrients and without differentiating external influences. Of course, the possibility cannot be excluded that only some [epithelial] tissues differentiate purely as a result of a bequeathed developmental capacity, while other tissues, perhaps [in large part] the supporting substances, are differentiated secondarily from the embryonic blastema by actions of what is bequeathed.

But we still know nothing of the manner by which processes of both kinds are possible, nor how they take place in their essence, for we observe only the course of external phenomena. Changes in full-grown humans take place only [?] through external, transformative influences, while embryonic differentiations take place without, or almost without, such differentiating stimuli. Nevertheless, it is reasonable to assume that embryonic differentiation is lawful, albeit this lawfulness is currently incomprehensible. The essence of embryonic differentiation, and its individual physical-chemical

---

1 [This sentence is faulty. Instead it should say: As a consequence of the continuity of the specifically structured germ-plasm, the sameness or similarity of children to parents is a mechanical necessity. This continuity is based on more or less perfect assimilation. Apart from disruptions, the parents themselves merely represent a previously developed part of the continuity. The mechanical necessity is also a consequence of the circumstance that "typical" ontogenesis takes place through self-differentiation of this germ-plasm.]

causes, are currently completely closed to us. Therefore, it would be without purpose to delve further into them at this time.

We are left with the question of the earlier phylogenetic causes of tissue differentiation. We shall have to be content with highly hypothetical statements because actual information for answering this question is very limited. We lack knowledge of the real causes of tissue differentiation in the past, but we still have an opportunity today to observe the emergence of some differentiations by definite causes.

Thus, after a bone fracture, in which the break is not adequately repaired, we find formation not only of bone but also of cartilage and connective tissue in the large inflammatory mass. The frequent rubbing of the fracture ends against each other leads to the development of a pseudo-arthrosis or "false joint." This formation of specific tissues under definite conditions from an indifferent disposition is a principle of greatest importance histologically and [*nota bene* if there is bequeathal of acquired properties] also in comparative anatomy. It allows us to accept discontinuous emergence of formations of the same tissue—for example, of the cartilaginous preformed skeletal parts—and it explains Adolf Eugen Fick's finding that the embryonic ribs of tritons are cartilaginous and, from the outset, separated from the axial skeleton; that is, they do not acquire their independence only by secondary formation.

Bone often forms in connective tissue in places that are frequently rubbed or hit, such as the so-called exercise bones and riding bones.

All such metamorphoses of tissues are very meaningful for us; for here we see real differentiations of one tissue from another, not according to bequeathal, as in the formation of a new eye from the stump of a snail's cutoff tentacle, but by external influences without the participation of typically localized bequeathal. But the tissue differentiations that can be ascertained in this way as a result of known (external) causes are limited to various forms of connective substance which are remodeled into each other.

Although nothing certain can be established today about earlier phylogenetic causes of tissue differentiations because these differentiations are currently bequeathed and we lack any understanding of embryonic self-differentiation, it nevertheless does not seem superfluous to offer a few further thoughts about this question.

The various tissues are affected by various functional stimuli, which can produce a chemical change in the cells of the tissues, be it an excitation that is connected with material turnover in the form of consumption, as in muscle, ganglionic, nervous, and sensory cells, or an excitation that is predominantly associated with excretion, as in the deposition of secretions by glands and deposition of intercellular matrix by supporting tissues.

It is now important for us to discuss whether the stimuli triggering these specific functions take part in original tissue differentiation (even if only by cultivating it) or whether that differentiation is due solely to Darwinian / Wallacian accidental variation of organisms. In the former case, a kind of self-organizing would be taking place. In the latter case, the preservation of organisms is due to how useful the variations are to an individual organism. Here, what was written on the struggle of parts in the second chapter comes to our aid, and we shall refer to it several times and repeat what was written there.

Suppose, for example, that living compounds arose by chance variation that reacted in some way to light in some cells on the surface of the body of lower eyeless animals, whether by absorbing light or converting it into heat by means of pigment granules, or otherwise being altered thereby. This would have been possible in three ways. Either the life process of a still indifferent cell, not specially adapted to and preserved by any other stimulus, was weakened by light in its regenerative capacity and assimilation; then it would have perished in the struggle of parts and been gradually eliminated, as we have shown above. Or second, the vitality of the association was unaltered by light, in which case the cell would linger. Or third, light strengthened assimilation; then the substance must achieve victory and spread, unless other equally powerful substances in the neighborhood were able to offer resistance.

However, it is important to consider that these three possibilities have different probabilities. The middle case, in which the substances are not in the least altered in their life force by light, is merely a special case from the middle of an infinite series of possibilities and as such, mathematically speaking, is extremely improbable, quite apart from the continual exchange of happenings. For just as an unstable balance is not found in nature as a permanent state, a substance can just as little endure unaltered in the exchange of all happenings unless it is maintained by special regulatory causes.

On the other hand, and on principle, the other two possibilities have the same probabilities of occurrence but not of maintenance. The former case, where light weakens assimilation, has no chance of being preserved, compared with the special case of no effect of light and, of course, still less compared with the third case in which light increases life activity. From this can be concluded that—although the origin, through a permanent or recurring living force, of assimilation substances that are unfavorably or favorably influenced is equally likely, *ceteris paribus*—only the favorably influenced variety can be preserved and thus also increased. In my opinion, these mathematical probabilities, combined with the exclusive possibility of preserving the stronger in the struggle of parts, gives these theoretical considerations a certain positive value, not merely a heuristic value.

In the present case it follows that in cells that are still indifferent it is more easily possible that variation occurs in processes that are altered by a stimulus. Of these variations, only those whose assimilation is strengthened by the stimulus are capable of being preserved and found in present states.

The struggle of parts is thus a principle of the cultivation of life processes (living substances) in organisms that are more and more strengthened by, and therefore react more and more strongly to, the living forces of surrounding nature.

The possibilities of such connections are, of course, extremely varied. The struggle for existence of individuals will, as explained above, pick out and hand over to definitive preservation only those variations that prove useful for the whole. In plants, for example, those connections will be preserved that consume light most perfectly. In animals, on the other hand, qualities [deserving the name visual substances] will be preserved that absorb light most perfectly in the cells of the retina, but consume it least and best prepare it for transmission to the brain so that the individual's eyesight becomes as sharp as possible. So it is by no means ruled out that, once different substances have appeared through variation that are excited by a stimulus, different qualities develop for the same stimulus and can be cultivated by it in an ever more extensive manner.

In the same manner, organisms must have adapted to all specific forms of the living forces of nature that occurred often enough or persistently enough for as long as this adaptation varied sufficiently—that is, as long as the organisms have not yet achieved a certain resistance to altered external circumstances through self-regulation, to which can be attributed the present rather considerable constancy of the kinds of beings living today from protists to humans.

Thus, it can be explained that organisms exist that have particular absorption organs for all specific forces that occur in nature more frequently: for light, warmth, sound, chemical movements, and movements of masses. And if electrical motions were more widespread in nature, lasted longer, and were not of too great an intensity, a particular kind of organ of perception would have been developed for them as well.

In order to prevent misunderstandings, of course, it must be mentioned parenthetically that the production of living forces by organisms (that is, the production of mass, heat, light, and electrical excitation) is something completely different from adaptation to an active living force and does not belong here.

Whether a special adaptation has been cultivated for every rate of vibration within a form of force—for example, for each frequency of light, aside from possible molecular-structural qualities—naturally depended on the benefit of such adaptation for the individual.

An individual's need was satisfied once it was able to perceive the entire spectrum of vibrations of a form of force with three different organs and to reduce them to these three components.

Anyhow, it must have been of great use to animals to have absorptive substances for all forms of force that flow through space and that are therefore able to mediate relationships between distant parts (in that each opposing part more or less absorbs and modifies the forms of force according to its own nature, and the force impresses upon the part recognizable signs of its presence). [When suitable variations occurred, sensory substances arose for all existing forms of force: recording sight, hearing, taste, smell, touch, and heat.] Therefore, it was a matter of course that the struggle of individuals extracted certain adaptations that had been cultivated by the struggle of parts and gradually cultivated them to an ever-higher perfection for the perception of external events. [On the other hand, the formation of specially structured sense organs built up from different tissues, represents more complex processes, in which the shaping effects of functional adaptation, to be discussed later, are engaged alongside the cultivation of formal variations.]

But because we have no organs for the perception of oscillations faster than ultraviolet light, although such organs would be useful, we can perhaps infer backward from this that such forms of force, which are theoretically possible, occur only very weakly or rarely, if at all. The cause of their absence could perhaps be attributed to the size of the molecules or to the tensions in the media that connect them that do not permit oscillations faster than about 800 trillion per second. But, with the existing elements of the earth, there is also the other possibility that organic material was unable to produce from these oscillations an agitation that could be transmitted by nerves or that organic substances let them pass unabsorbed.

Both the chemical and so-called mechanical forms of force (of motions of mass) vary greatly among themselves, and therefore the adaptations that exist for them are also very diverse.

The same physical precondition exists for the action of chemical forces, namely molecular contact, whether the substances coming into contact with one another are liquid or gaseous. Therefore, essentially only two forms of organs exist for chemical perception. But as the chemical qualities are different, so too are the adaptations to them. And even if we still have no understanding of how sensations occur, we know that we can sense thousands of different tastes and odors that cannot be broken down and regrouped into a smaller number of elemental sensations as can sounds and colors.

Most specific sensory elements have a receptacle for sensory excitation, the sensory hair, the origin and differentiation of which can be thought of in two ways, depending on whether it is a cuticular structure and withdrawn from material exchange, thus acting as it were purely mechanically, or whether it is a living structure chemically changed by excitation. In the case of cuticular formation, the sensory hair is merely an excretion of the sensory cell that is itself dead. In this case, the sensory hair must have acquired the ability to receive the sensory stimulus without the direct cultivating effect of the struggle of parts but, in accordance with Darwin, by selection from miscellaneous variation in the struggle for existence among individuals.

In the latter case of a living structure, which is more likely in our opinion, the substance of the sensory hair itself may be changed and strengthened by the sensory stimulus. Hence it unfolds proportionally to this strengthening and is regenerated, and perhaps also grows, independently through greater or lesser intake of food from its associated cell or from the environment. This last growth of the sensory hair will only be purposive within very narrow limits because too strong a growth would partly disturb the perception-capable shape of the sensory cell or entire sense organ. Therefore, it is understandable that processes have been cultivated in the sense organs that are capable of only minimal overcompensation, if at all. The last-mentioned type of origin of sensory hair would be simplest, namely the development by accidental variation of an appendage of the cell; and through the investigations of Wilhelm Kühne, we have also come to know phenomena of material exchange in the eye rods which are expressed as their swelling during activity.

As the sensory stimulus passes through the sensory cell, its quality will undergo a change, and it therefore does not seem striking that this new quality cultivates its own special organ, the next-following ganglion cell. Thus, through the originally existing ability of the sensory cell to overcompensate, which led to its proliferation, several cells may have arisen one behind the other, which are of different quality and gradually metamorphose the stimulus during passage in the manner necessary for the brain ganglion cells, as we can imagine for the three ganglion cell layers of the retina. Their nature is so different from the sensory cell that they, like the optic nerve itself, cannot be directly excited by light. Similar behavior recurs in other sense organs insofar as the excitability of the conducting parts for the specific stimulus is either entirely absent or is much less than that of the end-organ. Selection in the struggle of individuals as well as of parts will have been involved as much, or even more, in the circumstance that the stimulus does not pass through several ganglion cells in some sense organs. Indeed, both types of struggle must always participate in cultivating activity at the time that an innovation occurs, so each individual type of struggle always represents only one component of cultivation.

Just as the stimuli acting on the organism from outside could cultivate certain reactive substances in the struggle of parts, from which the struggle of individuals extracted only those useful to the whole, in the same way could the living forms of force and stimuli "produced" by the organism itself cultivate reactions substances, of which only the most useful were kept by selection of the whole: smooth and striated muscles, gland cells, and connective substances.

We would especially like to speak briefly of the chemical stimuli produced by the organism. Just as pilocarpine acts directly on the sweat glands to induce secretion, even after nerve transection, so too can chemical constituents of the blood stimulate the activity of the kidneys, pancreas, perhaps also the testes and liver. For the latter organ, the mechanism of regulation of the activity would be very simple if the digestive constituents supplied by the portal blood produced the stimulus. Perhaps the enlargement of the mammary gland during pregnancy is due to the stimulating action of chemical constituents, which stem from the child's material exchange. Moreover, the regulation of the activity of the lymph glands and spleen will, in my opinion, best be thought of as mediated directly by the composition of the blood.

The mechanical stimuli acting in the organism can then have developed various reactions of the organism. These mechanical stimuli that strive now to pull the parts apart and now to compress them are partly produced by muscular activity, partly by gravity, and partly transmitted from outside in other ways, and vary in intensity, magnitude of locomotion, duration, repetition, and angle of attack. For wherever a certain single combination of these properties occurs constantly, it will be able to cultivate a certain quality, as we see in those tissues that are exposed to purely mechanical stimuli in the case of the connective substances. A different stimulus will form bone from that which keeps the costal and articular cartilages alive and protects them from destruction and ossification. And likewise, the adhering connective tissue and the elastic fibers will have been cultivated by other stimuli in the struggle of parts. No hypotheses will be expressed here about the characteristic stimuli for each of these tissue qualities, but sooner or later they certainly must be attempted if the underlying conception is recognized that the functional stimulus is identical with the originally differentiating one or at least currently has a trophic effect. But very deep comparative-anatomical empirical knowledge will be required in order to correctly grasp the essential, common dynamics of the conditions under which each of these types of tissue occurs, even to be approximately correct, because the conditions in each individual are so different. The transition zones between divergent differentiations of the same blastema will be of great value here.

2. THE DEGREE OF adaptation of tissues or cells to a specific stimulus could differ, according to what has been set forth in the chapter on the struggle of parts. First, in such a way that the stimulus is able to strengthen assimilation: even without stimulus, the organic substances are able to regenerate and thus maintain themselves to some extent, just as we assume that without stimulus they steadily corrode, albeit more slowly.

Second, in such a way that the stimulus must be regarded as indispensable for the maintenance of the organic substances: the experiments on the muscles, glands, and nerves mentioned in the previous chapter showed such rapid degeneration of the parts after the withdrawal of stimuli that these tissues belong to this second kind. Furthermore, we know, from examples like Kaspar Hauser,[2] how insignificant our soul functions remain when the stimulation of them is neglected in youth, and our own experience teaches us how receptiveness is diminished by prolonged mental and sensory inactivity. From this it can be concluded that the functional stimulus is also indispensable for normal maintenance of the brain. We also had reason to assume that the matrices of the connective substances of the full-grown individual do not, physiologically, secrete any intercellular substance if they are not stimulated—that is, if they are not supplied with living force.

It seems, therefore, that the tissues of the higher animals require functional stimuli for their normal life in the same way as plants require light. Whether this also applies to lower animals cannot of course be judged without appropriate observations. But some facts—especially the high re-

---

2   Kaspar Hauser (1812–1833) was claimed to have grown up in total isolation in a darkened cell. —Trans.

generative capacity, which, according to earlier investigations and the most recent investigations of Paul Fraisse and Justus Carrière, is able to restore almost every cut or cutout part from the immediate vicinity in its typical manner—indicate that here, in lower animals, the cells are not fully adapted to their specific function but that each still contains a remnant of real embryonic matter [reserve idioplasm], be it in the nucleus or protoplasm, which comes into action as soon as, and to the extent that, it is no longer prevented from doing so by resistance from the physiological environment.

The more frequently a stimulus acts, the more perfect must be adaptation to the stimulus and, accordingly, the dependence upon the stimulus. And if a substance is accustomed to being excited daily or hourly, it will suffer more if the stimulus is absent for several days than will another substance that is accustomed to being stimulated only rarely. This adaptation to the "frequency" of the stimulus is a very important aspect. In the same way, there can be adaptation to a familiar middle "intensity" of the stimulus.

Bones that are used more often, such as bones of the extremities, are more easily subject to inactivity atrophy than less frequently used bones, such as bones of the skull.

Whether the different tissues originally came about during phylogenesis through embryonal variation or through some postembryonal influences, and whether in the case of the latter included a functional stimulus or not, in any event the tissue substances concerned came under the dominion of the latter (postembryonic influences) because we observe them (the tissues) as being dependent on the influences. The formal development of the parts, so perfectly adapted to function even in the lowest vertebrate animals, was brought about in this way. In the second chapter, we had reason to believe that the "functional stimulus" for many parts, especially for supporting organs, is indispensable for their present formation in embryonic and postembryonic life. But from this we do not get any point of reference for judging whether, in current embryonic development, the embryonic independence of the parts from stimuli ceases "by itself" because phylogeny repeats itself in ontogenesis through bequeathal, or whether this cessation of the independent ability of embryonic parts to sustain themselves takes place with the cultivation of stimulatory substances and is only conditioned "by" the action of the functional stimuli.

Whether one or the other is correct, it is understandable that some pathological bone formations (exostoses and the like), which already develop in the embryo or only develop later from remnants of embryonic substance, are capable of self-preservation because they neither represent a repetition of phylogenetic equivalents nor are they exposed to stimuli. Exostoses can remain unchanged for life on a bone, which, if it were inactive, would be subject to pronounced atrophy.

It is also understandable that parts of glands that have never been strongly active, which are perhaps only constricted covering epithelia, such as the hypophysis or brain appendage, the epiphysis or pineal gland, and the thyroid gland, even after their function has ceased—that is, without being hit by the surface stimulus as usual—remain permanently alive whereas other active glands completely atrophy within a few weeks after complete withdrawal of stimuli. [These examples are no longer appropriate because it has since been shown that the thyroid gland (and perhaps also the hypophysis) has an important secretory function and that the epiphysis represents the remainder of an eye primordium.]

So we become dependent on a stimulus through its effects, just as plants are dependent on light and cannot live normally without it. They develop as embryos in the dark, but they require light for later unfolding and normal growth.

Thus, in the full-grown individual, where there are no longer any embryonic properties apart from regenerative substances and tumor germs, the parts normally will only grow in response to stimuli, for they have become completely dependent on stimulus.

Complete adaptation to a stimulus would mean that every substance would be stimulated to function only by the stimulus that physiologically affects only it and could be kept alive and induced to increase by it alone. But adaptation has not progressed so perfectly in any tissue, for it is well-known that the nerves and muscles are excited by all living forces (even if, according to Paul Grützner, not to the same extent) except sound and light. Special experiments with different forms of force that measure the quantity of living force used or required for stimulation will reveal to us the various adaptations to particular stimuli.

To emphasize this, two periods must be distinguished in the normal life of all organs of higher organisms: (1) an *embryonic period in the broader sense,* in which the parts unfold, differentiate, and grow by themselves [without the functional stimulation directly or indirectly necessary for this], and (2) a period of *functional stimulus life,* in which growth and, in some parts, even the complete replacement of what is used up only takes place under the action of functional stimuli. The latter stimuli can then also produce something new, which [according to earlier assumption], if it had been generated in this way through generations, becomes hereditary; that is, it develops in offspring without these stimuli and becomes "embryonic" in our sense.

Likewise, a "gradual" decrease in the necessity of the life stimulus can also take place. For when the stimulus gradually decreases, other substances can be cultivated to adapt to the reduced frequency and intensity of the stimulus, and by this means organs can be preserved despite reduced activity, as we see in the human ear muscles, which are only weakly excited by central stimuli to an insufficient degree for contraction and yet remain intact if only to a very slight extent. Such preservation will only be possible where, as is the case for the ear muscles, the organ has no struggle for space. Elsewhere, where organs must fight for space, less-used organs can persist where they are strengthened by a small measure of function sufficient for robust resistance, as clearly shown by the rudimentary but active red *muscularis plantaris* of the human calf.

The time at which the period of "embryonic" life ends and "stimulus life" begins for each tissue and organ is probably different for each part. We have shown that the vessels, bones, and connective tissue probably do not develop their normal shape completely independently in the embryo. Indeed, this dependency is probably not just morphological, in that there is some morphological connection between the development of a muscle and its fascia, but a functional one in the sense that the "dynamic order" of the fiber course of the fascia is determined by the embryonic function of the muscles. The same is true of the blood vessels, which, besides the heart, are likely the earliest parts to begin their stimulus life. This is perhaps followed by the connective tissues, but no doubt in the various organs at different times. The general rule is that those organs that already perform their specific function in the embryo also lead a stimulus life in the embryo, according to the measure of this function. We do not generally know whether the glands already function, but we have reason to believe that the kidneys and liver do.

Humans only react to light several hours after birth and to sound considerably later according to William Preyer's observations. If some animals were not born already hearing and seeing, then one might believe that functional stimuli were first necessary before these sense organs could become functionally competent. Perhaps nerve tracts in the central organs must first be made passable by the stimulus. In any case, however, it does not seem to be undeveloped as a result of a lack of time. If forty weeks and a few days were necessary for adequate instruction, then a deficiency, lasting ten to twelve weeks, would be very evident in children born prematurely. Because this is not the case, it seems to me reasonable that these sense organs lack perfection only in their finest molecular proportions at the time of human birth, which only the functional stimulus can produce [but their visible structure is independent, that is to say here created and developed without the aid of functional stimuli].

B. Quantitative and "Formative" Effect of Functional Stimuli

After this consideration of the qualitative effects of stimuli, which is somewhat detailed for its necessarily hypothetical character, let us move on to their quantitative effects, better their formative effects, in order to get to know some of the properties that were not sufficiently discussed in the previous chapter when comparing the possible achievements of processes invigorated by stimuli with the actual arrangements of organs.

1. Size Relations in General

We have seen that the functional stimulus strengthens assimilation up to the point of overcompensation for what is consumed, and that therefore its strengthening effect must also increase with the strength or frequency of the stimulus, whereby a principle was given of the most purposive quantitative self-regulation of organ development. This self-regulation works in such a way that an organ becomes larger and stronger through greater use and thereby becomes capable of better performance. It also follows that an organ that needs a functional stimulus to assimilate reduces its nutrition and volume when it is less used, entailing a highly purposive saving of material.

But this happening is linked to material exchange; and it does not matter morphologically whether material decomposition is more or less linked to the function, as in the case of the muscles and glands, or takes place with a certain independence from it, as perhaps in the case of the supporting substances. In reality, we do not know anything about the material decomposition. Only for bones, Albert von Kölliker and Georg Wegner have taught us that particularly large cells, the osteoclasts or myeloplaxes, constantly dissolve the bone substance in many parts of the organ while at the same time new bone is formed in other places by other cells, the osteoblasts, so that a steady material exchange of the organ takes place; even if it does not take place within the cells, as in the working organs, it exists in the complete removal of larger parts, almost perceptible with the naked eye, and formation anew in their place. We also know that bones become weaker when inactive: the cortex being thinned from the inside and the individual trabeculae thinning and decreasing in number. One of the most striking examples of inactivity atrophy of the bones is the complete disappearance of the tooth processes of the jaws after the teeth have fallen out in old age by which, for example, the lower jaw is degraded by 1.5 to 2.0 centimeters in height and thereby reduced to a round brace the strength of a pencil. This atrophy can be explained in the same way as the atrophy of the working organs, in that in the absence of the functional stimulus no bone or less bone is newly formed, while the dissolution of bone either remains the same or increases. But nothing is known about the laws followed by the dissolution of bones in the struggle of osteoclasts against the bone substance, about which places are attacked more strongly, whether perhaps in places that are no longer affected by the stimulus or in those that have long acted.

We also have no knowledge of the material turnover or physiological regeneration of the connective tissue; but observed atrophies indicate a material turnover of the tissue; and it is perhaps most probable that the process takes place here as in the bones, that here the white blood cells dissolve the fibers in a normal manner, as they do pathologically in inflammation, while the diminished functional stimulus of the connective tissue cells induces the secretion of new fibers in an insufficient measure for replacement.

[To the comment of a contributor (Ernst Krause in *Kosmos*) that hypertrophies of the tissues are often detrimental, both those of activity and those of inactivity, I added the following: "Pure inactivity atrophies and activity hypertrophies are always more purposive for the prevailing performances of the organism. If, however, the neglect of the use of a part of the body and the consequent low development of it will one day be detrimental to its author, then it is not the mechanism of the organism but only the minute understanding or weak will of the individual that must be held responsible. The former would be the same as reprimanding a common soldier for not carrying out a necessary assignment that he had not been given by an officer. If, on the other hand—for example, in spinal polio—the ganglionic cells for certain muscle groups in the spinal cord have been destroyed by illness and the use of these muscles is thus abrogated, then the atrophy that occurs in the muscles in question and in their supporting organs, the bones, ligaments, etc., can be seen as quite purposive.

Cardiac hypertrophy with valvular defects has the character of the utmost purposiveness, for it enables the heart to cope with the increased resistance caused by the valvular defect and thus to continue the operation of the blood circulation under very difficult circumstances. But the fact that the heart also becomes hypertrophic over time, in the manner of functional adaptation, when it intensifies its activity purely nervously, in the case of nervous palpitations, can less be attributed to functional adaptation but falls under the point of view just described in the case of inactivity atrophy.

However, self-sufficient hypertrophies—such as the genuine muscle hypertrophy first demonstrated by Leopold Auerbach, which is always associated with reduced performance, or idiopathic atrophies of parts— are always based on a morbid quality that deviates from the one we represent that is cultivated in both instances of struggle; and the detrimental effects of these morbid qualities can therefore not diminish the purposive achievements of the qualities cultivated in these two modes of struggle."]

2. *Molding of Outer Shape*

If all size proportions are thus developed in a manner corresponding to physiological needs by way of self-regulation, the same happens with the shape proportions of organs, in many cases by the same principle.

If the stimulus is primarily located in a part of an organ—for example, in the case of a particular mode of excitation, on the lower or upper edge of a muscle, which consists of fibers in different directions, for example, the large breast muscle—the muscle fibers will only multiply at this point while atrophy perhaps takes place on the opposite edge through little use, whereby the whole shape of the organ undergoes change over the course of time. For such a lasting change in use, however, continuous causes of this change are also necessary. Such a continuous cause can be given by the will in a businesslike necessary mode of movement; but it can also consist in the fact that, through embryonic variation, the joint ends of a bone have undergone a change in shape that changes the mode of movement. And, vice versa, the same thing can take place in the bones if the muscular arrangement has been changed by embryonic variation; for the pressure of the muscles acting differently during the activity will accordingly reshape the ends of the joints. [For example, an angle joint can become a screw joint if an antagonist is shifted to one side from a pure confrontation. Strictly speaking, two organs that are functionally correlated will always adapt to one another, but depending on the magnitude of the variation and the faster or slower malleability of one or the other, the change will presently take place more in one or the other part.] The same fate must be shared with the associated joint ligaments, and the fascia must also be given a different structure in accordance with the modified pull. As a striking example of such a reshaping of bones, I recall the shape of the skeleton of a clubfoot: here all the bones of the tarsus and the metatarsus appeared considerably changed in accordance with the new conditions.

A similar change in the shape must have taken place in the brain through uneven use of its parts, when the specific elements laid down in particular parts were stimulated to proliferation through particularly heavy use, so the area in question was enlarged ahead of the others. Only here the process will have proceeded very slowly, so the change in shape became noticeable only after a change in use continued for many generations; while in the case of muscles and bones, the change occurs over the course of a few years, indeed it develops in a recognizable way in small animals within a few months insofar as a change of movement has been enforced by artificial or pathological alterations.

Whether an unequal distribution of function also takes place in the glandular organs is unknown, and it is probable only in the event that dissimilar qualities have previously arisen due to embryonic variation. I therefore believe that the change in shape of these organs in activity hypertrophy is primarily due to unequal external resistance.

3. Formation of Organ Structures

Not only the external shape but also the internal structure can be influenced by the same principle of strengthening by the stimulus and can be directly formed in the most purposive way, insofar as the functional stimulus itself has a definite shape or strives to assume such a shape.

This is most recognizable in those parts that have a static function because here the stimulus assumes certain forms that graphic statics teach us to recognize. Everyone knows that the pressure in a bent or obliquely loaded column is not evenly distributed over the entire cross section and that it propagates along certain lines. These lines are determined by the shape of the structure itself as well as by the position and shape of the surface to which the pressure is first transmitted. In the same way, pressure in bones must propagate most strongly in certain directions, and the bone-forming cells (osteoblasts) lying in these directions are therefore hit hardest by the stimulus for bone formation and are therefore most active in the formation of bone. From this it follows that these directions must become most prominent even if the resorption of the spongeous substance by osteoclasts is assumed to be evenly distributed. And it also happens that, if these directions are formed sufficiently firmly by bone substance, they remove the pressure from other directions so that no bone can be formed again at these points after resorption. Furthermore, if the osteoclasts interrupt the pressure lines, as may well be the case, the pressure will be distributed to other neighboring particles, and these will become stronger as a consequence of stronger stimuli. So also in the peculiar material exchange of bones, it goes hand in hand with the complete destruction of the formed parts. The structure corresponding to the static pressure lines must develop again and again, as actually happens, even in completely novel conditions such as with crookedly healed bone fractures.

The inner structural upheavals, shown by Richard Grossmann and Julius Wolff to be necessary for appositional bone growth, find their explanation in the same manner. These authors' objection to the theory of bone growth was that for the same bone during its growth to always maintain the same static structure of the spongiosa, continuous inner upheaval must occur with resorption and regeneration. The factors determining the execution of this remodeling, however, have hitherto been completely unknown. According to the principle of "functional self-organization" of the "functional" (specifically "static") structure presented here, it follows of its own accord that every bone, as it grows, re-creates essentially the same structure, but on a larger scale, by dissolution and reformation. The outer shape of the new bone remains "similar" to the old bone in the mathematical sense so long as the mode of loading does not change. And it is just as self-evident that when these conditions change, the structure corresponding to the new distribution of pressure must develop by itself.

In such a latter case, such as in the case of a broken bone, it is perhaps not superfluous to think through the whole process according to our conception. After the fracture of a bone (even one that does not damage the skin nor shatter the bone at the fracture ends), the osteoblasts of the inner and outer bone sheath (the endosteum and periosteum) and of the Haversian canals that run through the bone are continually subject to small insults of movement at the fracture site. They will be extremely sensitive to these insults because they have been accustomed to living in almost absolute calm, tightly clinging to the bone [that is, because they are normally protected from mechanical insults and are only struck by molecular vibrations when the intact bone is used]. Because mechanical stimuli have a trophically stimulating effect on osteoblasts, they begin a very impetuous proliferation, with defensive bone secretion protecting against movement that continues until protection is sufficient and peace is restored or possibly until the bone-forming force is exhausted, which not infrequently takes place in weakened individuals before the new consolidation is complete. [Insufficient bone formation is also said to occur in strong individuals in whom the plaster cast is so well applied that it completely ensures peace. However, this is possible only in rare circumstances.] Apart from the cast, peace is restored when the ends of the fracture are rejoined by a continuous, sufficiently thick bone mass.

Once the fracture has healed, the situation suddenly changes: the unfamiliar stimuli cease, and, once again, the only stimuli are static ones propagated into the newly formed reaction mass in certain directions due to the strain of the old bone parts. Thereafter, new formation after resorption will take place within these pressure lines, so the static structure corresponding to the new conditions gradually develops while the remaining callus mass and any protruding bone ends are more and more resorbed over the course of time. [This deduction is made with the minimum of assumptions. In reality, resorption will not take place just anywhere but mostly in unburdened locations, whereby the directions of greatest use must gradually be worked out in a meshwork alone. The combination of the two principles, applying force at points of greater pressure and absorbing at points of relief, will of course greatly accelerate the establishment of a functional structure. The resorption of the unburdened, however, seems to proceed much more slowly than the accumulation at places of strong pressure so that the latter thus has a greater share in the formation of the static structures.]

In a similar way the formation of static structure will have taken place in the tendons, aponeuroses, ligaments, fascia, and the eardrum. Those cells that are struck most by the stimulus in certain, fixed directions, by the pull, also secrete the greatest amount of intercellular substance; after sufficient deposition, fibers lying in other directions are completely deprived of the stimulus and cannot be formed again after their physiological shrinkage.

To go into detail, because the fascia and the eardrum are subjected to pulling in different directions, over the course of their generation the fibers develop only in the two directions that are most used and to which all others can be decomposed; for even if the fiber structure were originally disordered, the cells lying in these directions must become hypertrophic by stronger stimulation. At the same time, the life-sustaining stimulus is withdrawn from all oblique directions (according to the magnitude of the cosine of the angle), making their regeneration impossible. Two such components in a surface, standing in suitable directions to one another, will, if they are sufficiently strongly reinforced, completely relax all other directions; therefore, the directions of the two more heavily used components must ultimately remain the only ones substantiated in all two-dimensional structures, in that they vanquish all other directions by depriving them of stimuli in the struggle of parts. The same thing, which occurs in three dimensions in bones, occurs in two dimensions in connective-tissue sheaths.

This reduction of effects in many directions to the "strongest" (most heavily used) components is a highly purposive decomposition. I consider this to be the most important and indisputable evidence for the stimulus hypothesis that I have put forward, and I have therefore used it above in this way. It shows that something develops by itself that applied physics has recognized and described only recently. Its probatory power lies in the fact that, despite the relevant formations being infinitely varied in their specificities, they all find their most perfect explanation through this single proposed hypothesis.

How many generations were necessary to instruct such a perfect reduction to two components can of course only be judged when we have established through observations in novel, pathological conditions how great is the individual range of adaptation in this respect, and how great the bequeathal thereof.

However, it must not go unmentioned that, in soft formations of connective tissue, at least hints of fiber arrangements in constant directions of strongest pull could have arisen from a disordered arrangement simply by mechanical rearrangement due to repeated pulling.

Such a decomposition into components of two fixed directions could, of course, not take place in those connective-tissue organs that can be spoken of as used alternately in different directions, such as the skin and the joint capsules [the latter even on angular joints where external forces act at an angle]. In these cases, the system of fibers must remain disordered. If, however, some directions are preferred, fibers must preferentially develop in these directions, as we can see in the skin on the extensor side of joints.

The effect of stronger activity hypertrophy in the direction of more frequent use and the subsequent withdrawal of stimuli and inactivity atrophy in less-used directions is not limited to the development of internal structural details of organs but also extends to the development of the "layout" and "shape" of entire connective-tissue organs, and their products here again bear the character of the greatest purposiveness.

Let us imagine, for example, the urinary bladder as a small organ, newly created phylogenetically in the vertebrate series, which was initially attached to the anterior abdominal wall only by scant connective tissue. At first, no separation of fibers could be distinguished. If this organ persisted and grew for a long time, then gradually differentiations would have occurred in the homogeneous attaching layer of connective tissue, resulting from the fact that the tissue was pulled more strongly on some parts and in some places by the bladder and its contents; partly from the habitual posture of the animal; partly from directions dependent on the configuration of the surrounding structures. Just as before in the smaller relations within organs, as the attaching tissue hypertrophied in these most heavily used areas the surrounding and intervening tissue became more and more relaxed and accordingly atrophied. As soon as the favored parts were strong enough to withstand the pull by themselves, their surroundings became completely relaxed and completely atrophied so that the reinforced parts now appeared as discrete ligaments. This discrete separation would have become all the more pronounced, the more constant the directions of pull and the less the surroundings pulled in different, alternative directions. Thus, we see that the discrete separation from the surroundings of the accessory articular ligaments are so sharply developed that they have an almost shiny surface whereas this is, of course, not the case with the ligaments attached to the urinary bladder because of the frequently changing direction of pull. A more uniform direction of fibers develops with a greater constancy of the direction of pull.

So this same principle of the trophic effect of the functional stimulus in the struggle of parts in connective tissues leads, in addition to the development of the most purposive internal structure, to the development of the strongest, more-discrete organs in locations of highest performance.[3]

---

3 [The functional adaptation of the connective tissue is almost as important as that of the skeletal parts and muscles. Many women's diseases are based on insufficient development of the connective tissue of the pelvic floor (fascia pelvis and other parts attaching the uterus). This aplasia stems from insufficient "pelvic floor gymnastics" during youth. Such gymnastics

But I do not mean to claim that "all" ligaments "arise" in this way by functional self-organization. Rather, their construction will have taken place many times through embryonic variation and cultivation according to Darwin while their finer form and continuous fiber direction will have been developed only secondarily by functional adaptation. This seems to me to have been the case for the *ligamenta coracoacromiale sacrospinosum* and *sacrotuberosum.*

Only the most thorough comparative anatomical investigation can decide these questions. Because we have demonstrated the possibility of functional self-organization, we will not hesitate to invoke this ability when such investigations show us the initial conditions were such that the observed organization could have been brought about in this manner.

Larger proportional relationships in bones also arise on the same principles than from the static arrangement of the spongiosa. Because outer parts of columns supporting greater loads have to offer more resistance as a result of their tendency to bend, the outer parts of bone will be stimulated to thicken by activity hypertrophy. However, to the extent that the bone thickens outward on all sides as a result of bending stress, the inner parts in the middle of the bone become unencumbered so that, finally, a complete loss of bone substance occurs within the bone through inactivity atrophy, resulting in the formation of hollow tubes. There is thus a principle of bending stress of always thickening the bone outward and, when bending it in all directions, also hollowing it out inside. By this means, the greatest possible strength is achieved with less and less bone substance; for the larger the diameter of a hollow column, the thinner its outer wall needs to be. Now even if we do not know why the external growth of the long bones of this kind does not continue indefinitely but rather comes to a definite end, earlier in mammals and later in birds, the process itself must be traced back to the causes that have been given here. And there is no reason not to allow that what we have said of the tubular bones also provides in the same way a dynamic explanation for the cavities of the frontal bone, upper

---

are inadvertently practiced in the poorer classes by hard physical labor. In the more affluent classes only mountain climbing, ice skating, playing lawn tennis, and the like serve in a similar manner. All activities that increase pressure in the abdominal cavity have a beneficial effect in this sense if they take place frequently but not too intensively all at once. . . .]

jawbone, sphenoid bone, ethmoid labyrinth, and mammillary process of the temporal bone, even if here too the cause of the eventual external de-limitation of the process remains unknown.

Which tissue occupies the internal free space after atrophy—whether bone marrow is formed as in the tubular bones, or whether adjacent epithelia grow back to line the space as in the cavities of the cranial bones, or whether as occurs in birds the cavity is lined by outgrowth of the lungs—will at any rate be determined by accessory aspects, the explanation of which nobody will ask for at this point.

On the way to self-organization under the influence of stimuli, cavities no doubt likely arise in the bursa and tendon sheaths where great displacements of neighboring organs occur against each other due to overstretching and subsequent atrophy of the loose connective tissue. On the other hand, the origin of the pleuroperitoneal cavity, and even more of the subdural area around it, can likely be traced back to embryonic variation. Not only the shrinkage but also the quantitative development of the loose connective tissue of the *perimysium externum,* which everywhere corresponds exactly to the degree of displacement between neighboring muscles, can thus be understood as being brought about by functional self-organization, because they are always precisely individually adapted to circumstances as they are conditioned by occupational activity and the like. They are needed by and therefore cannot be assumed to have arisen by arbitrary variation and selection of the purposive (as according to Darwin).

As I have described, the form of the lumen of blood vessels, which presents the shape of a jet of liquid freely emerging at a branching point from the oval lateral opening of a tube through which there is flow, must also be understood to result from formative stimuli. I showed that this fine characteristic shaping of form can arise only when the blood vessel wall has the wonderful ability to resist strong side pressures but to give way completely to every impact of jets of liquid, even the immeasurably finest, that is to yield to every one-sided pressure. This is especially true of the intima, the innermost skin of the blood vessel wall, which has no nourishing vessels of its own; for this reason alone these nourishing vessels could not have contributed to the creation of its arrangement. There can be no question here of a mechanical self-organization by the liquid jet—that is to say, of a passive reshaping of the wall material by the forces of the flowing liquid—because it is impossible that a substance that can withstand a pressure of several hundred grams in certain directions without yielding in the slightest should yield to a pressure of a few milligrams in the perpendicular direction. For this we must appeal to the properties of living matter; but if we assume this property or quality, which is currently incomprehensible—as we do not yet understand the organic qualities at all—then all of the extremely useful various designs of the blood vessels in all parts of the body, mentioned in the first chapter, result by themselves so long as the branching itself is externally given.

In the case of the working organs (muscles, glands, ganglion cells, nerves) we do not know with certainty anything about the particular form in which the stimulus tends to disseminate, and which it therefore tends to impart to the structures in which it disseminates. But one might guess that it is for such a reason that the nerve fibers are cylindrical, of the same thickness in their whole undivided course, round in cross section and as straight as possible, never meandering, so that bends only occur when they are forced by external conditions. Even chemical processes, if they have a direction, follow the law of inertia and do not change direction without a particular cause. But why the sympathetic fibers are ribbon-shaped, we cannot deduce. Likewise, one might assume that the spherical or spindle-shaped form of ganglion cells, with a conical transition from and to the nerves, is due to the form of the propagation of excitation. But given our ignorance of the circumstances, other conjectures can just as easily be made.

But the stimulus seems to have a direct formative influence with regard to the role of ganglion cells in the orderly arrangements (or coordinations) of thoughts and movements. According to the present-day view of physiology, we imagine that the arrangement of individual psychological impressions to form thoughts and the coordination of muscle fibers and individual muscle groups to form movements are mediated by the thread-like connections of ganglion cells, themselves the seat of the individual innervations. In the rich inborn network of threads between ganglion cells, the stimulating impulse of will makes some threads more passable and thus brings the ganglion cells and their functions into a more firmly established relationship, so the functions proceed more easily at the same time or sequentially. That is the way we must presently imagine the action of learning through repetition, as far as it occurs in the central organs and in the building up of coordination.

The formative effect on muscles is somewhat more evident, the least so for transversely striated muscles. Because in these latter, as mentioned in the previous chapter, the transverse stripes become indistinct after the nerve belonging to the muscle has been cut, it seems that the stimulus also has a polarizing effect on the disdiaclasts (flesh prisms in the muscle fiber), and that it thus maintains their ordering in transverse and longitudinal rows. The stimulus can also have a determining effect on other proportions of the fiber; because, however, I have begun a special investigation into this, I will forgo any further information at this point.

In the case of structures composed of smooth-muscle fibers, their form is definitely related to the effect of the functional stimulus as well as to the function itself. In order to explain the related forms, it must be assumed that in order to maintain "smooth muscles," not only is the functional stimulus necessary but also the "completion of function," the active overcoming of resistance with shortening. This assumption will probably not be disputed by anatomists because every anatomist knows that muscles diminish wherever this opportunity has disappeared due to a change in development. Jakob Henle says, "It is a well-known anatomical fact that muscles degenerate into connective tissue when the parts between which they are stretched become immobilized."

As is well-known, smooth-muscle fibers have no definite points of origin or attachment that give the fibers definite directions, but rather they form skins in which fibers could lie mixed together in random directions. But this is not the case. As mentioned in the first chapter, in the most varied of organs in which they occur, smooth-muscle fibers always lie only in the directions of strongest performance, and in this, once more, is expressed a reduction from many directions to the strongest components. Thus, we saw in the cylindrical hollow organs (such as the intestines, ureters, vas deferens, and blood vessels) only transverse and longitudinal muscle fibers, the origin of which deviates in the following way from the similar relationships of the connective tissue organs discussed previously. From a disorderly original organization, those fibers that lay in directions of the activity of the parts found the greatest opportunity for shortening and overcoming resistance, and accordingly deprived the other fibers that were placed at an angle of the opportunity to act. For the bladder-formed organs, such as the urinary bladder and gallbladder, which only present a locus of least resistance in the particular direction of the drainage outlet, toward which the greatest shortening is possible, we have fiber strands that meridionally overlay the organ from this location. Because this direction, which is privileged by function, is determined by the discharge outlet, it deprives all the fiber strands lying obliquely to it of proper material support by the cosine of the oblique angle. So only the fibers at a 90 degree angle, in concentric circles covering the organ, do the most work (apart from the fibers in the discharge outlet direction). For these reasons fibers are preferentially formed in these two directions, while fibers in all other directions become slack and succumb to shrinkage.

In the case of the uterus, which contracts relatively seldom in humans, we can perhaps trace back the less perfect arrangement to this circumstance of the rarity of its function, apart from the changes brought about by the lateral confluence of two channels.[4] In mammals such as pigs, rabbits, and mice, which frequently use their uterus [and whose young are also relatively small], the fiber arrangement corresponds in a high degree to our rules. At this point I would like to add that I am inclined to consider that the rapid atrophy of the muscle substance of the enlarged uterus that occurs after the child or a large tumor has been expelled, which reduces the organ's weight by two-thirds in a fortnight, can be interpreted as the result of the relaxation that has come about. For, if in this organ an enlargement of the contents gives rise to hypertrophy by tension, then complete relaxation after it has been emptied can be considered a sufficient cause of atrophy. In any case, I do not believe that atrophy is merely a consequence of the sudden anemia that follows expulsion because the cause of the spastic narrowing of the vessels that is necessary for this anemia would be unintelligible, and without spasm of the vascular muscles such a profound reduction in the blood supply cannot be inferred for hemodynamic reasons. On the contrary, the tension in the blood column will strive to fill the available paths as much as possible, both when the uterus relaxes and when it contracts.

---

4 Roux is probably referring to the embryonic fusion of the two uterine horns of the human simplex uterus. —Trans.

As far as the glands are concerned, we are entirely ignorant of any formative effect on these organs of functional stimuli. This is all the more difficult for me to admit because the question of the cause of the internal form of one of these organs, the liver, first gave rise to those reflections, the results of which I have presented to the reader in this book. This was the question of the cause of the peculiar conduct of liver cells, described by Ewald Hering and Albert von Kölliker, in which the tubular arrangement present in all other vertebrate classes is transformed into a cross-trussed arrangement in mammals. I believe, however, that in spite of our present incapability, the principles I have established will one day lead to an explanation of this difficult morphological problem once the ontogenetic and phylogenetic modes of origin have been explored in more detail, although the excellent work of Carl Toldt and Emil Zuckerkandl will be an essential point of departure. Perhaps I am permitted at this point to request the possible sending of pieces of fresh liver of the lowest mammalian forms in absolute alcohol, or soaked in Müller's solution, and to assure the donors in advance of my gratitude and my willingness for any possible counterservice.

Finally, when describing the formative effects of functional stimuli, it must be mentioned that the so-called dimensional activity hypertrophy, the exclusive enlargement in activity hypertrophy of the dimensions of organs that determine the strength of the function, should at least be included here.

C. Temporal Relations of Functional Self-organization

We are currently unable to judge, for the most part, the periods of time within which the self-organization of the described relationships took place under the influence of functional stimuli, and it is possible that hundreds or thousands of generations contributed to some formations [at least to the extent that formations resulting from functional adaptation are bequeathable]. Only in the case of bone tissue did we see that these structures can develop in a clearly recognizable manner within an individual's lifetime. The periods of time required are in any case very different for the various tissues, so it will perhaps have taken an incomparably longer time for the dynamic arrangement of the smooth-muscle fibers to have developed than the described structure of tendon membranes. [The changes, which Darwin summarizes under the name "correlative variability" as far as they directly achieve the purposive (that is the lasting), turn out to be functional adaptations if examined appropriately, in that they are based on the fact that, with the primary change of one part, the function of another part is changed secondarily, which now leads to a corresponding change in the shape and structure of this second part as well, as has been explained above. If, therefore, the principle of correlative variability is to be subordinated to the principle of functional adaptation in its directly creative effects, then the same applies in part to the reactions of the individual to external influences which directly produce purposiveness.]

On the basis of the foregoing, one might suppose that I am of the opinion that basically all formations have arisen through self-organization under the influence of a functional stimulus and must be kept alive by it. But, if all form comes about through the stimulus, it would still remain to be explained from whence comes the formed (and thus at the same time formative) stimuli.

It was already emphasized earlier, when considering the qualitative effect of the stimulus, that the parts under the control of the stimulus can only have come afterward through the lasting or repeated action of the stimulus and perhaps still come about in this way in ontogenesis. The consequence is that parts that are not subject to such stimuli, or only rarely subject to them, cannot be ascribed any dependence on stimuli. Experience shows that human adaptability, our ability to learn and become accustomed to influences, is greatest in youth and decreases qualitatively and quantitatively with increasing age. At the same time, our so-called regenerative capacity, which is already small in itself, grows ever weaker in old age.

These phenomena find their perfect explanation in our conception of the life of parts. Because, under the influence of stimuli, there is a cultivation of the stimulus substances and stimulus forms while embryonic indifference and the capacity of the parts to independently maintain themselves are progressively lost. The organism becomes more and more perfectly adapted to the stimuli over a longer period of time, becomes more differentiated and therefore more stable, so subsequent transformation into new properties and forms faces ever greater obstacles; for it is easier for something that is indifferent to develop into a one-sided constitution with the loss of its versatility than it is for something decidedly differentiated (one-sidedly constituted) to transform itself into something differently constituted. Because, furthermore, the development of stimulus life is connected with the loss of embryonic self-sufficiency of proliferation, the ability to regenerate is also progressively reduced, which I intend to elucidate more precisely in an experimental work.

It was discussed earlier that those tissue differentiations, which the ancestors probably first experienced through certain stimuli, can now arise in the embryo without the stimulus, and that most differentiations probably do so. What applies to tissue differentiation must also apply to "formal" differentiation. Normal properties originally acquired through the functional adaptation of adults develop in the embryo without this functional stimulus, and can be cultivated more or less completely in the youthful period without such activity, or with a minimum of it, and can be maintained for a while as bequeathed properties without the stimulus. But gradually they will atrophy if they fail to function, and in the course of generations they will become weaker and weaker in individuals, and also in bequeathal, and finally fade away.[5]

From this it follows that embryonic substances that are formed in excess can retain their embryonic property of self-sufficient growth (as was assumed by Julius Cohnheim for tumor germs). This is because these embryonic substances are not sufficiently reactive to functional stimuli to be brought into absolute dependence on them, either because they are protected from the action of functional stimuli by their aberrant position, or, if this is not the case, because they are not sufficiently reactive to later-acting functional stimuli on account of their developmental lagging behind.

Thus, embryonic epithelial cells that may have formed excessively or that have otherwise been accidentally pinched off from the epithelial surface can preserve their embryonic properties by being at a distance from

---

5 [The dubious principle of inheritance of the characteristics acquired by the individual is used here. The same results can, however, have arisen without this principle, albeit in another way.]

the surface and from the action of the surface stimulus. And it is conceivable that even substances that are not excessively formed, if they are protected from functional stimuli [or from other differentiating influences], retain their embryonic properties as a result of the absent connection (for example, embryonic cartilage or bone parts that have been relaxed or unburdened by improperly developed forms in their neighborhood).

One more difference that must necessarily be present in the origin of changes by embryonic variation and by functional adaptation must be emphasized.

The transformations of form that arise by functional adaptation on the path of changing use are only possible gradually and only in certain directions from the point of departure of the change. So, for example, the inner joint ligaments (the cruciate ligaments of the knee and the teres ligament of the hip) if acquired through functional adaptation (as is probable for the latter ligament according to the investigations of Hermann Welcker) could only arise through the gradual, and very specific, inward development of the joint capsule to allow the arrangement of muscle to move the joint and provide the functional stimulus. Their present complete self-sufficiency would therefore only be secondary, acquired through further changes in the external conditions of the muscular apparatus. [These changes must therefore have taken place in a manner similar to the changes that are made in a house or a bridge during continuous use. Because these structures must remain functional at all times, changes must take place in small steps in a definite sequence. By contrast, something completely different can be built in the place of an earlier structure that has been demolished.]

By contrast, and as far as we can presently judge, changes due to embryonic variation (which arise not due to the functional stimulus but due to minimal changes in structural qualities [of germ-plasm] or other aspects unknown to us) can produce new forms in any direction and from any starting point. So from this type of variation could be produced, for example, a teres ligament located in the middle of the joint and completely free from the wall, as well as a whole new muscle, such as an *adductor longus* in the forearm or extra fingers with all their accessories.

But if such embryonic variations have arisen, when the time comes for the parts to be used, the parts will alter their function, and the change in function thus forced will result in a corresponding remodeling of the parts in the manner described above. If, for example, a joint head has been changed by embryonic variation, the muscles will have to be used differently: some groups will develop more strongly, and others will more or less succumb to inactivity atrophy; and the same can result from embryonic changes in the ligaments. Or conversely, as already mentioned above, the passive parts, bones, and ligaments can be remodeled by embryonic changes in the muscles. Which of the two occurrences (adaptation of the passive parts to the active parts or vice versa) is more frequent we cannot judge with certainty at present. But I am inclined, in general, to ascribe a preponderance to the active parts leading the passive parts in this respect [?]. A part changed by embryonic variation will, with changes in its function, also alter the function of other parts and thus cause their corresponding transformation.

[This adaptation will, of course, be mutual and will last until both organs correspond to one another, whereby the more adaptable organ experiences a greater change from its original structure than the less easily adaptable one. This creates angle joints where a single antagonist group of muscles acts, but if the antagonists do not work entirely in the same plane, then a corresponding shape is created of a real or fake screw joint, depending on the circumstances. Ball joints are only formed where the muscles form at least two groups of antagonists of almost equal strength acting at right angles to each other; if rotators act at the same time, the spherical shape is even more ensured; if one of these muscle groups is almost completely stretched or paralyzed in a juvenile animal, the ball-and-socket joint originally present is transformed into an ellipsoid joint during further development.]

Thus, through the principle of trophic stimulus, the necessary *functional harmony* in the structure and function of the various parts of the organism will develop by itself, even when new variations appear. How quickly this happens and how much of it may already take place in the embryo can only be determined through special individual observations. The possibility of spontaneously harmonious development of novel characters that appear during embryonic life must be decidedly accepted for those structures that already function in embryos (namely, the blood vessels and, according to William Preyer, many striated muscles, and thus also ganglion cells and supporting substances).[6]

But there are also parts of the body that have no active or passive function, that are useful and have been preserved only because of their presence, their visibility to the outside world. Of this kind are many characters of sexual selection: the enormous dorsal crest of male tritons that grows only at the time of rut and regresses after the rutting season; the rooster's crest or the turkey's wattles have no active function. The form of these characters is therefore the result of embryonic variation, as is often also the case with the color, and probably always the case with the markings, of animals. But even if the whole organ as such has no function, the parts have a function in the organ—namely, the function of maintaining the whole organ. As one part comes under greater tension than another, an unequal function of the parts will develop within the whole, and with it a corresponding inner structure of the whole, thus in the aforementioned examples a static structure.

The same is true of begetting organs that act through their external form. Here the form was certainly created solely by embryonic variation.[7] But the internal arrangements show that the individual components have developed according to the measure in which they contribute to the production of this form. It is the same with other parts of the sexual organs. The

---

6. [In the foregoing, an innumerable number of individual different configurations corresponding to the concept of the so-called purposeful, in their form and structural relationships corresponding to this character, were derived from just "two" formative principles: from the exercise of the function and from the trophic effect of the functional stimuli. . . .

Because neither of the principles used here for explanation has an inexplicable teleological character, neither the fulfillment of the function of the organs nor the promotion of assimilation through the functional stimulus has an inexplicable teleological explanation. This is because all have a mechanical explanation. Therefore in this sense the explanation is mechanical, not teleological.

What is true for the "morphological functional adaptation" treated here also applies to the "purely functional adaptation" that mediates it; that is, for the quantitative and qualitative self-regulative adaptation in the respective "exercise" of the function, this too only has a mechanical character.]

7. The sexual organs are formed before they are used and therefore cannot have been shaped by functional adaptation. —Trans.

whole transformation through which, for example, the fallopian tubes have been separated from the ureters can only be traced back to embryonic variation and selection summed across individuals, according to Darwin, not to direct functional adaptation. On the other hand, as shown above, the structure of the longitudinal and circular muscles of the walls of the fallopian tubes can only be a consequence of functional adaptation.

The accessory apparatuses of the sense organs also belong here; for only the specific parts can be influenced by the stimulus itself, while the accessory apparatuses are all formed by embryonic variation and only determined in their structure and finer form by functional self-organization.

Embryonic variation is thus free to shape the parts outwardly in any arbitrary way; but then the inner structure of these parts, the arrangement that produces this outward form, is no longer free but adopts the most purposive arrangement by functional self-organization, possibly aided by the struggle of parts.

If, on the other hand, the outward form itself is exposed to certain influences, like the shape of the bones and ligaments to the effects of muscles, then the outward form is no longer free as long as the determining character of the other organ, in this case the muscles, is given.

Because what happens in the embryo is initially a purely chemical formation by chemical processes, it follows of its own accord that chemical alterations will be able to influence and change the formation of entire organ systems all at once. It thereby becomes easier to bridge large gaps in the animal kingdom, such as between reptiles and birds or between amphibians and mammals (as Alexander Graf von Kayserling has already pointed out). A chemical alteration, aided by functional adaptation, can bring about such a great change in the form of an organ system, or in all parts of the organism at once, as would perhaps not have arisen in thousands of generations by functional adaptation alone. A striking botanical example is described by Wilhelm Knop: the entire inflorescence of maize plants was transformed, with a change in the flowers themselves, after hyposulfuric magnesia was substituted for sulfuric magnesia in the diet. Most plants no longer formed the characteristic shape of a corncob. Only on the shortest plants did the tips of the husk of a corncob emerge from one of the lower leaf sheaths. Albert von Kölliker mentions another very interesting example: if the air supply to the chicken embryo in the incubated egg is absent or insufficient, then endothelial vesicles develop in the vascular area with many nuclei and endogenous budding and lead to the formation of blood vessels in a way that is entirely different from normal processes (these were erroneously described as normal occurrences by Edward Klein).

As we have seen, much of the formation of form depends on the action of stimuli, already partly in the embryo and even more in the adult. Furthermore, pathology teaches us that, apart from functional stimuli, the tissue also reacts plastically to other external stimuli. It directly follows that forms will be changed and must deviate from normal when tissues are subjected to foreign stimuli. One of the simplest examples is the development of congenital flatfeet, which, according to William Martin, Alfred Volkmann, Albert Lücke, Otto Küstner, and others, arises when there is an absolute or relative lack of amniotic fluid and consequently direct pressure of the uterus on the infant parts.

But development usually takes place in a normal or typical manner. This indicates that the organism is highly protected against abnormal stimuli, rather than against normal stimuli; and that the stimuli forming the normal form are produced in the embryo itself without external influences.

If artificial hyperemia of a part is produced in a young individual—that is, if more blood is supplied to the part than it is able to procure for itself by means of the above-mentioned self-regulation according to its bequeathed properties—then abnormally strong growth occurs with the formation of abnormal forms because the parts can still grow without functioning at this stage. Even in the full-grown individual, we must recognize in some tissues (the covering epithelia and supporting tissues) the ability to grow stronger simply as a result of artificially increased food intake. Every doctor knows the marked thickening of the bones of the tibia after violent mechanical insults (whereby the osteoblasts themselves are also irritated), [and it is astonishing how long the productive effect persists here after the initial insult,] as well as the increase in connective tissue in chronic inflammation. These formations, however, are not permanent but gradually disappear again according to the measure and speed of the material back-and-forth to which the tissue in question is subjected. A restitution of what has disappeared after the cause of its origin has ceased cannot take place, unless the formation has become stable and self-sustaining over the many-yeared endurance of the cause [?]. Incidentally, here too, as above, we must remember that we mostly do not know, even in these tissues, whether the hyperemia brought about by the stimulus was the sole cause of tissue proliferation.

Because functional stimuli produce so much that is purposive, one more word must be said about centralization of the functional stimuli of the whole individual, because the development of parts that are purposive for the whole depends on this centralization. The impulses of the will that emanate from the brain go through the ganglia and nerves to the muscles and thus, in addition to influencing the formation of these parts, also influence in a quantitative manner the formation of their supporting organs (neuroglia, tendons, bones, cartilage, ligaments, fascia). Stimuli that act on the body and the digestive organs from inner surfaces are also subject to the centralized self-regulation of the whole, and the same also applies, but only in a more imperfect way, to stimuli that strike outer surfaces and sensory organs of the body, to which the organism must furthermore compulsorily adapt because the center of will controls the introduction of substances into the body through its control of the organs of movement. Although all these forms are mediated and brought forth mechanically, they are in part truly responding teleologically to a purposeful will. However, this property does not exclude that they can also be detrimental to the whole if the will, as in the case of overexertion or refusal to eat, intends something detrimental to its bearer.

Thus, through the cultivating effect of the struggle of parts and through the stimulus life that thereby is victorious, a perfection of organization can arise along the way, which until a few years ago was hardly suspected and which we still do not fully understand in detail. When the adaptation of an organ's tissues to functional stimuli has been perfected, the organ has been educated to the most abstract but material definition of its function. It has been adapted to this function down to the smallest functioning particle. Every single organ, and the compound whole that is composed of the organs, sustains a perfection that, in our own human works, we can construct only in theory but never achieve in practice. A purposiveness of arrangements arises, which the struggle for existence among individuals (Darwin's and Wallace's aggregating and intensifying principle) could never have brought about by itself and which has only become possible through continuous combined effect of the struggle of individuals and struggle of parts.

The next tasks of research will be to identify such a possible perfection of the parts, up to the material definition of their functions for the individual they belong to, and more and more to identify this for each of the organs and tissues. This is especially needed for the almost completely neglected functions of the various connective substances.[8]

[All these qualitatively differentiating as well as formative achievements that are a necessary consequence of cultivation in the struggle of parts are initially only of value to the individual in which they occur.

However, to the extent that the origin of these assimilative qualities is based in the germ-plasm, the variations so cultivated can be transferred to offspring as a result of the continuity of the germ-plasm and its ability to assimilate. Because these processes of cultivation and functional shaping are thereby repeated in the development of each individual, the resulting properties can benefit the whole species. The great spread of the new qualities within individuals caused by the struggle of parts, and of the germinal variations on which they are based, greatly increases their possible advantage and thus their cultivability [*Züchtbarkeit*]. This is augmented by a considerable increase in these favorable properties through their cultivation in the struggle of individuals for existence because those individuals have been preserved who developed these more favorable properties to a greater extent according to the nature of the germ-plasm from which they arose and of which they still bear a remnant for their own reproduction.

All this happens without operation of the dubious principle of the bequeathal of "acquired somatogenic" properties of individuals. However, without operation of this principle, the higher degree of adaptation acquired through the struggle of parts in the organism during somatic development would always be lost with the death of each individual.

If, however, an individual is able to bequeath its "acquired" properties, then this latter large part of useful forms could be transferred directly to

---

8 Chapter IV of the first edition ends at this point. The remainder of the chapter (enclosed in square brackets) was added to the second edition. The major import of the extra material is two-fold. First, Roux is now less confident of the bequeathal of characters acquired by somatic cells (following the critique of August Weismann). Second, Roux's concept of embryonic development has shifted toward ascribing a greater role to evolution (or preformation) as opposed to epigenesis. In the first edition, Roux had accepted Ernst Haeckel's thesis that protoplasm (plasson) was homogeneous and amorphous. —Trans.

204 ~: THE STRUGGLE OF PARTS

the offspring, according to the extent of their bequeathal. The share of the struggle of parts (developed in this book) in the development of the entire realm of organisms would then be considerably greater than that of Darwin's struggle for existence among individuals, and the functional adaptation based on the struggle of parts would then be much greater than possible without the bequeathal of "acquired" characteristics. The struggle of parts (or selection of parts) in the individual would then be a much more important principle for the development of the whole realm of organisms than struggle and selection among individuals.[9]

After I had presented (in the first edition of this treatise) the theoretical possibility of action of the three principles of the selection of parts, of functional adaptation, and to the eventual bequeathability of their actions, my task was then to determine their actual magnitude of action. Since then, I have tackled this task with regard to the first two principles. With regard to the selection of parts, however, it soon became apparent that our knowledge of the physiologically changing inequalities of cells of the same tissue and of cell parts (of inequalities that cannot be transferred to the offspring of cells) was still inadequate to be able to look forward to success in an investigation of the cultivation of lasting, transferable qualities.

The theoretical question of whether acquired properties can be bequeathed has been excellently discussed in the works of August Weismann, Johannes Orth, Heinrich Ernst Ziegler, Friedrich Rohde, and others; and some authors have already tackled it experimentally, so we can hope that gradually more certainty will come to our knowledge in this extremely important but also extremely difficult area.

I have begun to determine the actual scope of functional adaptation of the individual's form and shape in a few contributions on the morphology of functional adaptation with regard to the performances of connective tissue, muscle, and bone tissue. This has also been undertaken for bones

---

9 [If there could be any personal interest in scientific endeavors, then as the founder of the doctrine of the purposive shaping actions of the struggle of parts, I would have the greatest interest of all descent theorists in the proof and recognition of the "bequeathal of somatogenic properties." However, I consider Weismann's doctrine of the continuity of the germ-plasm and the bequeathal of variations in its substance alone, with the elimination of the principle of bequeathal of somatogenic properties, as such a great relief for our cognitive faculties that I urgently wish this doctrine to turn out to be true.]

by Julius Wolff; for muscles by Hans Strasser and Hermann Nothnagel; for glands by Nothnagel, Hugo Ribbert and by Conrad Eckhard and others; and for vessels by Richard Thoma and Nothnagel. It is noteworthy, however, that most of these authors only undertook their work as specialists without knowledge of the other principles represented in this book and therefore without any connection to them.

Some of these works are neither experimental nor do they address pathological facts (experiments of nature) but draw conclusions from normal formations and processes. They therefore only provide indirect demonstrations, of which one must always remember, however, that they can never guarantee complete confidence on their own but first need additional experimental proof.

In this treatise, functional adaptation has been used in a double sense:

- first, as the derivation principle of the so-called direct development of the purposive;
- second, as a principle of the ontogenetic derivation of a large number of "typical" individual forms from a small number of bequeathed characters, by this means reducing the share of typical forms that are directly bequeathed or (expressed in other terminology and using these terms in the deepened senses that I have introduced) by this means reducing the share of evolution and increasing the share of epigenesis.[10]

In general, if we review the empirically obtained results on the scope of the functional adaptation to novel conditions, its assumed significance has been confirmed with regard to direct adaptation of bone, muscle, glandular tissue, and the vasculature. No experimental work has yet been published on the related behavior of the connective tissue, but I have undertaken such a study which will be presented in greater detail in the relevant special work. In this work, I have discovered that the role of functional adaptation in the formation of functional structures during postembryonic life is

---

10 Roux here opposes evolution, in its older developmental sense of the unfolding of preformed characters, to epigenesis, the emergence of characters from what is unformed. He argues that most characters develop by epigenesis from a few inherited (preformed) characters. Roux treats "functional adaptation" as synonymous with "epigenesis." —Trans.

reduced in connective tissue relative to other tissues. Inactivity atrophy proceeds only very slowly, so fibers that are no longer used are preserved for a very long time, thus impairing the purity of the functional arrangements, in directions of greatest stress, of what is formed by the pronounced activity hypertrophy.

As for the empirical share of functional adaptation in "typical" ontogeny, it seems to be considerably less than would have been possible with the most sparing application of the principle of evolution; that is to say, some ontogenetic formations arise independently of functional adaptation, even though they could easily have been produced by functional adaptation with only a few given formative factors, and despite the presence of these factors. I merely remind here of the rather extensive development of the first vascular system in the absence of an embryo in hens' eggs (and, according to Jacques Loeb, in embryos with paralyzed hearts), the independent laying out of the forms of joints before they are used, etc. Many individual investigations will therefore be required in this area in order to ascertain the true extent of the role of functional adaptation in the development of typical forms.

Finally, it should be mentioned that some authors have raised objections to functional adaptation. However, such objections do not have any serious significance because it is evident from the explanations that these authors did not delve into the area sufficiently to be able to make well-founded judgments.[11]]

---

11 [This concerns statements by Wilhelm Pfitzner and Bernhard Solger. Pfitzner found that strong individuals sometimes have poorly developed muscle ridges on the bones of hands and feet, while female persons with flaccid, poorly developed muscles possess sharp ridges and strong insertions; he therefore believes that the increased development of all those muscular and ligamentous attachments is more or less pathological. . . . On the basis of these observations, Pfitzner, who "never believed in functional adaptation," indulges in hints that the modern view of the importance of the function for the formation and maintenance of the skeletal parts is misguided. He asks, "Can function create a thing out of nothing?" but does not indicate who held such a position. . . . Solger objects to the functional significance of bone structure; I have discussed his objections elsewhere.]

CHAPTER V

On the Essence of the Organic

In the infinite variety of natural happenings we know of a kind of process, or, to express ourselves more commonly, of formed beings in which processes take place, that differs so clearly from all other happenings by a sum of properties that, from earlier times, it led to the classification of all being and becoming as either organic or inorganic. In spite of this, however, it was not possible to clearly grasp and define the real nature of these processes, even if every age has tried to do so.

[A few definitions by prominent recent authors can at least be cited here:

Herbert Spencer defined life as "the definite combination of heterogeneous changes, both simultaneous and successive" and finally offered the following formulation as the most general and perfect definition of life: "The continuous adjustment of internal relations to external relations." This definition is so very "general" that the concrete content of life has almost evaporated and no one obtains clarity.

Much more suggestive is Haeckel's definition: "Because organisms exist without organs, we must abandon the morphological definition that living bodies are natural bodies, that is, consist of various devices that serve the whole, and justify the concept of organism on a physiological basis and accordingly name organisms as all those natural bodies that exhibit the peculiar activities of 'life' and especially that of nutrition. On the other hand, we call inorganic all those natural bodies that never exercise the functions of nutrition or any of the other specific 'life activities' (reproduction, voluntary movement, sensation)." I have not listed "growth" because inorganic individuals (crystals) also grow, nor reproduction, because reproduction has been lost in many (asexual) organic individuals.]

Depending on one's point of view, depending on the natural scientific knowledge one possesses, one's judgment must be different and more or less close to the truth. That is why Aristotle, the greatest natural scientist of antiquity, had one of the best definitions: he recognized that, in organic beings, every part had specific functions; it was a tool, οργανον, for the whole; and therefore he called the whole an "organism," meaning a complex of tools. Since, however, we have come to know living beings without special organs—that is, beings that merely represent a continuum of similarly appearing substance—it has become doubtful whether this definition denotes the essence of the organic or only an outstanding property of higher organisms; and philosophers have already withdrawn their approval because these substances seem to lack the "inwardness" or an encompassing soul.

Let us see whether, from the standpoint of the present, we are able to advance the question a little farther, to come a little closer to the essence of the organic.

The "connection of many visibly different parts into a whole" cannot thus determine the essence of organisms because there are living beings without this diversity of parts. [This distinction therefore only connects to the vulgar concept of the "different" parts of the organs as macroscopically and microscopically visible different parts. Strictly speaking, however, correspondingly different parts must also be present in every smallest particle of the protoplasm (isoplasm) capable of assimilation.]

Just as little can the psychic functions of organisms comprise what is essential, for we have no well-founded reason to ascribe psychic functions to the lowest animals and plants. As far as we know, these living beings perform all their functions without consciousness. Just as little can "mechanical memory" (the persistence of the cause in the effect) serve as the characteristic feature, for according to the law of inertia it is a general function of matter or, more correctly, a property of everything that happens.

"Being-for-itself"[1] is also not to be mentioned here because this applies just as much to every process separable from the environment due to its consistency or other special qualities—just as much or, more correctly, just as little. Strictly speaking, being-for-itself does not exist anywhere but is merely a more solid interconnectedness and interaction within a thing than with its environment, and its degree results from the nature of the distinction from the environment and the interconnections within the thing. [In organisms, the connection and nature of parts is such that all parts work together to maintain and preserve the whole in its particularity, and because the whole consists of nothing but insubstantiations of processes, these processes endure longer, last longer, than if the parts did not work together.]

Furthermore, the essence is neither the ingestion and consumption of living force nor the conversion of tightening force [*Spannkraft*, tension]; for both types of force exchange occur continuously in inorganic happenings; and just as little is it material exchange combined with force exchange, because that combination is shown daily in evaporation from the surface of water, in weathering of rocks, etc.

Nor does a certain consistency or physical composition form the essence of living material, even though, for active parts, these fluctuate only within narrow limits. But there are no living materials of the same soft, colloidal quality, nor can one derive life phenomena from these properties alone. Accordingly, these properties can only be regarded as favorable, perhaps necessary, preconditions. The same is true of concentration, which varies from 12 percent water in legumes to 98 percent water, combined with 2 percent solids, in the organic matter of jellyfish.

Perhaps a certain collective chemical combination is what is essential, for fluctuations in this relation are not very great, but this is probably not the essence itself because the chemical constitutions of plant and animal protoplasm are very different with almost opposite modes of chemical action.

But after exclusion of these properties, only a few remain that must be drawn into the narrower circle of consideration. First of all, sensitivity or

---

1 Mechanical memory [*mechanische Gedächtnis*] and being-for-itself [*Für-Sich-Sein*] are Hegelian terms. —Trans.

irritability is an essential characteristic shared by all living beings, if not all their parts. It is the ability of organic structures to change shape under the influence of living forces, the so-called stimulus, in a manner that cannot be viewed as simply passive transformation through external influence but as "excitation"—that is, the triggering of a certain process within matter, associated with the consumption of stored energy, which increases cohesion or removes internal resistance to cohesion.

This reaction in the form of "reflex movement" is only a special case of the general reactivity of all substances. There are inorganic substances, for example, a mixture of chlorine and hydrogen, that by adding small amounts of living forces are strengthened by chemical reaction in their cohesion and condense into bodies occupying less space; but reflex movement is so different from all inorganic reactions that it can be accepted as a characteristic attribute of the living.

Reflex movement, however, does not suffice by itself for characterization. No one will, if the other properties are ignored, describe a form with this property as "living"; and we can also imagine organic processes with material exchange, growth, and definite shape that lack this property; nothing proves to us that this property is essential for life. But with this we are already anticipating what is to follow. Sensitivity can be described as a peculiar secondary property but, as we shall see, a very useful one.

Let us now proceed to the examination of the behavior of life processes, of the spatial and temporal behavior, in the *a priori* properties of everything that happens. We shall discuss spatial behavior first: the expansion of the living. Here we encounter the important properties of growth and reproduction.

Growth is, however, not an independent property but merely designates the quantitative reaction of another property, assimilation, and must therefore be regarded as dependent on assimilation. As is well-known, growth, considered as simple increase, also occurs in inorganic things, such as in crystals, and also in the expansion of a process from a smaller to a larger space, such as the excitation of air by insolation or evaporation, or the formation of mist, etc. Reproduction, what Haeckel has termed "growth beyond the individual level," is similarly dependent on the property of assimilation.

But the temporal behavior of organic processes is of great importance. The organic processes, as far as we can presently judge, have been of uninterrupted duration since their first coming into being. We are compelled to assume their continuity from the beginning. However, there are also inorganic processes that, like organic processes, have been eternally and uninterruptedly continuous from the beginning, and that have changed only in their intensity and extent since their inception—eternal weathering of rocks, eternal waves of the sea, eternal evaporation of water, eternal shining of the sun since its creation.

This demonstrates that eternal enduring, continuity of becoming, does not in itself reflect the essence of the organic, and yet this enduring is absolutely necessary. For we know that once the continuity of living substance is interrupted, nothing can restore it, because the thread has been broken. No scientist today questions that higher organisms have been continuously differentiated from lower organisms, simpler and simplest. Thus, organic processes must have endured and been lasting. This uninterrupted "lastingness" is the indispensable precondition of the organic, but it does not distinguish organic from inorganic processes.

[This fact is signified by the fundamental theorems:

Omne vivum ex ovo (Harvey)
Omnis cellula et cellula (Virchow)
Omnis nucleus e nucleo (Flemming).][2]

We shall have to investigate by which property the lastingness of inorganic and organic processes is guaranteed.

Inorganic processes are physical and chemical. Enduring inorganic processes are associated with place exchange or material exchange or force exchange; because process signifies alteration, that is change. They are

---

2   The three aphorisms were added to the second edition although Roux did not mark this with his usual square-brackets. We have inserted square-brackets. Roux did not give citations: *Omne vivum ex ovo* [Every living thing from an egg] comes from William Harvey (1651); *Omnis cellula et cellula* [Every cell from a cell] appears in Rudolf Virchow (1859, 25); *Omnis nucleus e nucleo* [Every nucleus from a nucleus] appears in Walther Flemming (1880, 363; 1882, 367) —Trans.

therefore associated with material consumption or force consumption or place consumption or multiple kinds of consumption simultaneously.[3] A cannonball consumes new falling space or flying space in the steady progress of its motion. Space has been consumed once the path has been flown.

Inorganic processes "with material exchange and force exchange" continue just as long, and only as long, as the external conditions by which they are generated continue; but as soon as they are no longer generated by these external conditions, inorganic processes end. So weathering proceeds as long as the atmospherics (air, carbonic acid, water) touch the rock and ends with the ending of their coming together; and it starts again when they come together again because weathering is conditioned solely by these external factors. The inorganic process is nothing for itself, but merely the consequence of the spatial and temporal coming together and acting together of external components. It is usually not considered as in any way being for itself. Consequently, it is difficult for the inexperienced to place side by side in their imagination an inorganic process that, for example, erodes the surface of a rock and organic processes that take place in discrete beings.

The life process is otherwise. Its conditions are not merely external, but, on the contrary, it is something for itself, not "merely" dependent on external conditions. If we combine the external preconditions of organisms—for example, the foodstuffs of plants and sunlight—or if we do the same with the foodstuffs of animals, no organic processes arise from their combination. It is only when these preconditions are introduced into an organic process itself that the life process is thereby increased. The life process thus bears the cause of its preservation within itself, and nourishment is just the precondition, whereas inorganic processes require just these external conditions in order to immediately come into being.

Thus, the organic processes have an additional condition to fulfill; and it might seem that therefore they would be even less lasting than the inorganic processes, but the result is exactly the opposite. Life processes are more lasting. We see them [or at least their specifically structured sub-

---

3 *Ortswechsel, Stoffwechsel, Kraftwechsel* are juxtaposed with *Ortsverbrauch, Stoffverbrauch, Kraftverbrauch.* —Trans.

stratum] in eternal continuity, despite changes in some of their conditions. They must have special properties that enable them to last; if we seek these out, we must come to the essential properties of organic events, to the features that distinguish them from the inorganic.

The first property that favors the persistence of organic beings under unfavorable circumstances is their ability to assimilate. Organic processes are able to transform externally procured parts into what is the same as themselves; to regroup atomic and molecular groupings that are different into new groups that are the same as themselves. When only the raw materials are available, organic processes are able to procure things that are qualitatively foreign and from them produce what is permanently necessary for organic persistence. The essence of assimilation is thus a kind of self-production, "self-organization of what is necessary for preservation, for continuance." In this, there is already an essential advantage over inorganic processes.

[The faculty of assimilation enables the formation, by living substance, of the same substance from foreign nonliving substance; it enables the transfer of the law of inertia from simple physical processes to life processes. It forms the basis of bequeathal: that is, the transfer of properties from cell part to cell part, from cell to cell, and from compound individual to its offspring, in the latter case mediated by the continuity of the germ-plasm.] But one inorganic process also has the property of assimilation and yet is unable to sustain itself permanently. "The flame" also assimilates foreign material.

Various possibilities may arise in the degree of the capacity for assimilation, the lastingness of which is different and therefore important for our investigation. Either the process assimilates less than it consumes, so it soon must cease of its own accord. This quality excludes lastingness. Or the process assimilates no more than it consumes; then it will never go beyond the extent of its coming into being. And if the conditions change at its place of origin or its current abode, if there is a lack of food or if external disturbing elements arise, it will be destroyed. Given the constant change in natural occurrences, it can safely be assumed that such changes in circumstances will occur. Only such assimilation processes can therefore be lasting that assimilate more than they consume. If this occurs to a sufficient extent that the processes can spread more and more, over larger areas, then the probability of their preservation despite change of external conditions increases correspondingly. For even if the greater part is destroyed, somewhere a part will be preserved.

Thus, after assimilation, growth—"overcompensation" in the replacement of what is consumed—is the next general requirement of organic beings. As is well-known, all organisms have this ability, which can be defined dynamically even if we do not know how it comes about. Overcompensation consists in the fact that more assimilative forces are set free in the course of the organic process than are necessary for mere replacement of what is consumed; or, conversely, that the transfer of foreign material into the organism requires less force than the assimilated materials supply in completion of the process, and that this surplus force promotes further assimilation.

The flame, once again, offers the simplest and therefore the most understandable example of assimilation. A flame frequently shows its capacity to assimilate more than it consumes by reaching around in a terrible way. Nevertheless, it does not possess eternal earthly lastingness. This is not its fault; on the contrary, its lastingness is very great and, as is well-known, often withstands the impact of the best steam-powered firehose. The reason for its perishing is mostly consumption of its material. Visible combustion in nature would probably have eternal duration if it did not proceed faster than other natural processes are able to create new material. In the organic, on the other hand, there are two kinds of opposing life processes, connected with oxidation and reduction, which, with the self-elimination of the unsuited, have settled into a balance that guarantees eternal duration.

Some processes may assimilate more than they consume but nevertheless not use everything they consume for assimilation, so that energy remains to "achieve" something, as we are accustomed to express ourselves. Here, we treat assimilation as merely a precondition for the latter event, the "achievement." Thus, in addition to overcompensation in assimilation, a flame produces light and gives off heat to the environment. But this achievement contributes nothing to the flame's preservation. Light is of no use to the flame. It is a loss, a waste for assimilation and thus for lastingness. Such processes must therefore, *ceteris paribus,* be inferior to those that use a greater proportion, or all, of the forces to increase their lastingness.

But this latter need not simply be done in such a way that everything is used directly for assimilation. It can also take place in the form of *achievements* that benefit lastingness—for example, when an achievement, such as the ability of a moneran to move, increases its ability to acquire food. The moneran expands its feeding area by stretching out parts of its body; then when it contracts immediately on something touching a projecting part, it takes in more food than had it simply remained as a compact ball. Contractility also accelerates digestion, in that the parts that are admitted internally become better mixed; by this means, homogenization is not solely dependent on the slow action of diffusion. Free locomotion also provides the great advantage of being able to leave regions of depleted food.[4]

An achievement that has been preserved because it benefits the whole by contributing to its lastingness is called a "function" that is "performed for the whole." The formation of light is indeed an "achievement" of the flame or, more correctly, of combustion, but it is not a "function" in this biological sense for it is of no use to the flame. It is a mere useless expense. It would be best for the flame if it were a pure assimilation process in which the flame's necessary fuel for survival, to wit, its nutrition, is a given, and the flame did not produce more than it assimilated. Instead the flame uselessly consumes its food material at breakneck speed and, in this, already lags behind organic processes.

From earlier times, many have continued to regard every process that takes place in a "part" and that benefits a more or less complicated "whole" as something wonderful. However, this usefulness for the endurance of the whole is by no means the "intention" of the parts. The parts live only for their own preservation; that something useful for the preservation of the whole happens is conditioned solely by the fact that only such properties could be left, and are left, behind; whereas other properties, achievements of parts, that were at least a million times more numerous occurred without being of any use to the whole and ruined the whole, and thus excluded themselves from lasting with the whole. But it is probably unnecessary to explain again the effectiveness of Darwin's principle. It is enough to re-

---

4 For Haeckel, monerans were simple forms of life composed of homogeneous, structureless matter. —Trans.

member that all life we now see is the summation of all "self-preserving" remnants of earthly happenings before our time. All processes that were not lasting in themselves or, despite this inner ability, not also lasting in their external relations, simply ceased; and we only find traces of their earlier activity or not even this; while everything that happened over the course of millions of years, in the eternal change of events, that by chance is lasting has been retained, just as our cultural acquisitions accumulate from the vast amount of transient ephemeral achievement.

The above achievement process of the moneran, its continuous rhythmic movement, is called automatic if it occurs without any particular external cause but is called reflexive if it takes place only in response to an external influence. From the outset, reflexive actions had the advantage over automatic actions of greater lastingness because environmental conditions are never constant. Thus, an achievement that proceeds uniformly cannot always have the same benefit; it will often be futile, often too modest, the latter when the external conditions are more favorable but are unable to influence the achievement.

On the other hand, reflexive achievements produce a highly favorable interaction with the external conditions they should exploit. If appropriate conditions are absent, then the achievement will also be absent. If appropriate conditions are present, the achievement results, and depending on the intensity of the external conditions, the appropriate intensity of achievement will be produced by itself. Thus, reflex activity is highly purposive; that is to say, the mechanism of "self-regulation" increases the lastingness of the structure. By contrast, an automatism is generally a nonpurposive arrangement: on the one hand, materials are wasted when demand is low; on the other hand, there is insufficient action when demand is high. Automatism will therefore only be of use under constant conditions, when circumstances and needs are constant, which is to say very seldom. And indeed, it occurs rarely and never in a pure form, as for example in ciliates or in cardiac ganglia. For even then automatism is regulated by external circumstances.

With achievement, the consumption rate becomes more important in material exchange. As long as a process was merely one of assimilation, so long as everything that was produced from the material was used in assimilation for overcompensation—that is, for growth—consumption was in reality just a favorable precondition for the individual's enlargement. With achievement, however, expenses arose that in themselves did not increase assimilation, although they consumed material. In this case, processes in which these functions consume more than could be replaced would have been unable to last. Therefore, only those living beings will be lasting in which the rate of assimilation achieves an economic balance between the material consumption for these functions and the magnitude of the indirect benefit from these functions. All other processes must perish and exclude themselves from the ranks of the living.

With achievement and the consequent consumption, a new requirement emerges that is of the greatest importance and dominates the whole organic process: "*self-regulation*" of all activities.

Because reflexive achievements must be the dominant kind but occur unevenly, the rate of consumption must also be uneven—sometimes increased, sometimes decreased—and the question now arises how assimilation is regulated. If assimilation continues evenly, there will now be a surplus, now a balance, now a strong-sustained function, when death or self-elimination occurs. Only such processes can therefore persist in which assimilation is dependent on the rate of consumption or on the stimulus which evokes the consumption. Thus, with greater consumption, the endeavor to take in food and the ability to assimilate it must be increased rather than weakened by diminution of the substance. "Lasting processes" must be hungry; this is to be understood not as a conscious sensation of hunger but as a stronger chemical affinity for food when more food is needed. Thus, food intake and assimilation must also be subject to "self-regulation" as can be seen in the simplest way in the flame's failure to endure.

Self-regulation must also apply to the elimination of what has been used up. If excretion were to occur invariably and uniformly, then with greater consumption there would be an accumulation of altered substances that differed from the organism. In the best case, if the products of excretion were simply useless, their presence would at least inhibit the organism; but if they were not chemically indifferent, they would directly disrupt the life processes. So excretion must also be subject to "self-regulation" by need, for which the flame again provides the simplest example. The faster the flame consumes, the more heat it generates and the more it assimilates, but also the faster it diminishes the specific weight of its fuel, and the faster it discharges the end products of material exchange.

Self-evidently, just as with processes of pure assimilation, only those processes bound to achievement can be preserved that are accompanied by overcompensation, for the reasons given above.

The dependence of assimilation on turnover can occur in one of two ways: either the rate of assimilation depends directly on the stimulus, in which case the stimulus directly increases assimilation; or this dependence is indirect, in which case the products of material exchange that have accumulated in response to the stimulus stimulate assimilation.

Whether the dependence of assimilation on the stimulus is direct or indirect, what is important for us here is the degree of this dependence.

Assimilation can proceed quietly during inactivity, whereas during and after activity it can be temporarily increased. This kind of process is very preservable and, I believe, very widespread. In the lower animals, it is perhaps the general or dominant one [whereas in the higher animals it occurs normally only in the "embryonic period" or in parts that have remained abnormally "embryonic" and form tumors]. These processes can be identified by the fact that they are able to withstand greater exertion but continue to assimilate during prolonged rest and are thus not subject to inactivity atrophy. The elimination of such parts that have become superfluous for the individual can only take place by the slow means of selection among individual variation, as described by Darwin.

If, on the other hand, the process is such that the stimulus has become an indispensable life stimulus, and that without its influence, not only the achievement but also the assimilation does not proceed properly, then this process will only have a chance of preservation if the stimulus acts very frequently, when the strengthening occurs constantly so that the overcompensation after the activity is large enough to last a long time, even during rest. It will also be necessary that weaker stimuli, which occur more frequently, are also able to excite assimilation. However, in the permanent absence of the stimulus, inactivity atrophy will occur as a result of the lack of excitation of assimilation, resulting in insufficient replacement of what is gradually used up without exercise of the function.

This kind of process is thus bound to more particular conditions of existence, different from those of the previous one, and will therefore be of more limited occurrence in the whole series of animals and possibly also in the individual organism. But it has properties that provide a great advantage in the struggle for existence. It represents the most perfect "self-regulation" of performance while at the same time the greatest economy of materials, in that those parts that are needed are strengthened and enlarged according to the measure of their use, but those that are no longer needed regress, and the material for their maintenance is spared. This kind of process thus represents the highest economy, with highest performance of the whole, but at the expense of the independence of the parts, which has here completely ceased. These parts live only from the function they perform for the whole. They are like public servants who have gradually and entirely become servants of the public without interests of their own, who are completely absorbed in their service and can no longer live without it, but atrophy immediately after retirement, as is so often the case with old public servants. And one need not be content to say that these parts are "as such officials," but also the other way around: such officials are processes that are perfectly adapted to a single occupation. According to the explanations of the current treatise, parts performing functions perfectly adapted for the whole are a feature of humankind in general in almost all their parts.

Such relations are found only in the higher organisms and form their characteristic feature relative to the lower organisms, in which the parts can live, and live for themselves, without functional stimuli. [This distinction is the justification of the well-known sentence "the higher animals eat to live, the lower animals live to eat and grow." Or, in other words, the lower organisms are eating and growing machines, and the higher organisms are working machines. With regard to this distinction, however, individual higher animals repeat the phylogenetic sequence of stages in their development, in that they behave like lower animals during the embryonic period.]

At the same time, as mentioned, the conditions under which these qualities of stimulus life develop are such that they can be preserved and expanded by self-cultivation in the struggle of parts only in occupations where the stimulus acts often enough; while in occupations where the stimulus seldom acts they must succumb in the struggle of parts, even if suitable variations occasionally appear. But the struggle of individuals will strive as mightily as possible to preserve the whole, as a consequence of the highest performance for the whole with a minimum of material consumption.

In the chapter on the stimulus effect, I showed that there is reason to assume such a direct dependence on functional stimuli of the entire life processes of human cells: for muscles, glands, and probably also for the sense organs, and to a more limited extent for the nerves and ganglion cells. And the fact that in these organs, as a result of complete withdrawal of stimuli, there is no slow atrophy due to insufficient replacement but rather rapid degeneration of what is present, speaks for the direct life-preserving effect of the functional stimulus. Further, we saw that in the case of the supporting tissue, connective tissue, and bone tissue that even if there is not exactly the relationship of inactivity atrophy being caused by degeneration due to inactivity (which is less plausible for the intercellular substances), nevertheless stimulation strengthens the cells' assimilation and their separation from supporting substances. For this is the only way to explain that the structure of these parts comes to correspond to the form of stimulation.[5]

All the qualities that have been proven to be the only lasting qualities in the foregoing are at the same time also those that, once a trace of them has appeared, will be victorious in the struggle of parts within the tissue in question and thus attain sole control, as has been demonstrated in the chapter on the struggle of parts; so the spread of these most useful properties, once a trace appears, must be rapid by their double victory in the struggle of parts and the struggle of individuals.

---

5 [Small pieces of bone or bone-forming periosteum can be transferred to another organism with success and permanent preservation of the bone despite this dependence on the stimuli that are determined by the brain, provided they come to places where they can again encounter adequate stimuli. It follows from this that even this highest centralization of the parts of the organism is not mystical, individual, but merely mechanical, granted that food and stimulus (the conditions of existence of the parts), are present. Wherever a part of an organism finds its accustomed conditions, it can survive, no matter upon which individual. . . . There are no living conditions for the parts of an organism that belong to the individual as such; there are therefore no "individuals" in the strict sense, no "indivisibles," rather only persons.]

If we look back over the overall character of all the properties most essential for life, they are "self-organization" of what is necessary for preservation (or "self-regulation") and at the same time overcompensation in the replacement of what is consumed. Assimilation, overcompensation in assimilation above the consumption rate, and "self-regulation" in all activities are therefore the basic properties and necessary preconditions of life. No matter how complicated the processes may have become in the course of further differentiation, these properties must be preserved and must appear again everywhere in all novel formations, for they alone guarantee "lastingness" under changing conditions.

The "ability to self-regulate" can be more or less great, depending on the constancy or variability of conditions; and overcompensation in replacement can be limited to a certain period of life and then cease both for the individual tissues and in the formation of sexual products. At least, they remain the most necessary and most characteristic properties of everything organic, the essential preconditions of the organic. The increase in the number of these properties in many connected ways, and their development to the point of greatest economy, form the first essential property of the organic. The ability to contract only comes second [?], and form-shaping by chemical processes comes third [?].

[The aforementioned "form-shaping by chemical processes" is already an erroneous requirement because (as argued above) organic forms are not shaped by chemical processes but are created by chemical processes from what is already specifically formed. In other respects, too, this last version of the essence of the organic given above is not good and does not even fully embrace what has been presented above. We would better say:

Alongside assimilation as the basic property, two capabilities developed to the highest usefulness for self-preservation, form the essential characteristics of all living things. These are the increase in the number of "self-regulations" in the performance of all achievements of organisms, and overcompensation (at least temporary) in the replacement of what is consumed in these achievements, both in the transitory "purely functional" and the enduring "morphological functional adaptation."[6] Some of the

---

6   Roux is probably distinguishing between (physiological) adaptation during individual ontogeny and evolutionary adaptation during phylogeny. —Trans.

various "achievements" of organisms are available to every living being, at least temporarily: reflex movements (or apparent self-movement) and self-division. These special achievements can therefore be included as part of the essence of the organic. The general occurrence of another kind of achievement, the psychological functions,[7] has not yet been demonstrated, so for the time being these cannot be counted as part of the essence of the organic; conversely, however, this function, where it occurs, is to be understood as the highest organic achievement.]

The inorganic is preserved only through external conditions and, when these change, immediately ceases in its previous nature. On the contrary, structures that must (as organisms must) preserve themselves through change in external conditions must be able to regulate themselves in order to survive, and this continues because their other properties, once destroyed, are too complex to be created in a short time anew by chance and then be cultivated to higher degrees. If such a structure wants to continue quietly under changed relations, it simply perishes. This is nothing new but on the contrary a fact that is all too well-known, too often experienced; and it is as true of the parts as it is of the whole, how all basic conditions and properties apply equally to the parts as to the whole, for the whole consists only of the parts. Each must be able to "adapt" to the circumstances, and this is only possible through "self-regulation," in which the changed circumstances trigger useful functions for the whole.

As a result of constantly changing external conditions, "self-regulation" is the precondition, the essence of "self-preservation," preservation by one's own powers. Self-preservation reaches its limit at the limit of self-regulation.

It is beyond the scope of our work to enumerate here all forms of self-regulation that arose in the course of the later, higher differentiation of the animal kingdom. Eduard Pflüger compiled a number of them a few years ago and pointed out the fact of their general occurrence, without, however, recognizing or expressing their importance for the origin and characterization of the organic.

---

7   *Seelischen* ("soulish") translated as psychological. —Trans.

Pflüger set out the following general law: "The cause of every need of a living being is at the same time the cause of the satisfaction of the need"; and added two laws of special behavior: "If the need only applies to a certain organ, then this organ alone procures satisfaction"; and "If the same need applies to many organs at the same time, then very often only one organ procures satisfaction for all."

After that, Pflüger was certainly close to recognizing self-regulation as an essential property of the organic because it alone guarantees duration; but instead of saying this, he closes with resignation: "How this teleological mechanism came about remains one of the highest and darkest problems." I hope, however, that through the demonstration of those properties that alone can win victory and thus endure in the double struggle, this obscurity has been somewhat cleared, at least with regard to the principal aspects of origination.

It was a hindrance to Pflüger that he considered self-regulation to be a ready-made, innate mechanism, even though he had already followed the right path with a reference to behavior in pathological cases. We are not the "music boxes with a thousand or million songs, which are calculated and set in millions of possible ways in the course of life," to which he compares us.[8] Rather, we are devices that can learn new songs every day. We are like a quick wit, who immediately grasps the essentials in every situation and expresses it in a clever and pointed manner rather than a mere colporteur of jokes who selects the most suitable jest for the occasion from his accumulated supply; or like a genuine doctor who chooses a prescription for each case of illness according to the individual circumstances rather

---

8 Roux refers to the following passage in Eduard Friedrich Wilhelm Pflüger, "Die teleologische Mechanik der lebendigen Natur," *Archiv für die gesamte Physiologie des Menschen und der Tiere* 15 (1877): 57–103: "Perhaps a simile makes the mechanics of the appropriate reactions of living beings most clear. Think of a large musical music box, the interior of which is designed in such a way that, after its clockwork has been opened, it can play a thousand different songs. There are a thousand spigots on the box, which is connected to the mechanics of the box in such a way that a displacement of a single specific spigot caused by the finger of the hand triggers a very specific song. We can also think of the devices in such a way that the size of the displacement of the triggering spigot depends on the strength of the tones forming the melody. Now the different melodies represent the various processes in the animal's body that are necessary in order to satisfy the needs that may become necessary in the ordinary course of life or to compensate for disturbances that occur more frequently." —Trans.

than an old hack who sells to the sick public every day the same fifty prescriptions he has learned by heart. In the same manner, the animalic organism differs from one that is "self-controlled."

The latter expression is the correct designation for the conception on which Pflüger's work is based. Self-control is a kind of self-regulation that is set up for a certain range of variation on both sides of an immovable central point; but organisms possess self-regulation of a most general character, in which, after remaining in a deviant position for some time, this position becomes the new central point for a new range of variation. If the deviation always continues on one side, then the new central point can lie far outside the original maximum range of variation. This distinction is not as pedantic and superfluous as it might seem; it must even be emphatically emphasized because the former (self-regulation) is the basis of the progressive perfectibility inherent in organisms whereas the latter (self-control) merely represents a stability arranged for very many cases.

Let me now indulge in a word about the much-discussed problem of the origin of life—that is, about primal generation (autogeny, abiogenesis, spontaneous generation, heterogenesis, or other names). Of course, I am in danger of acting against my own convictions. For I am of the opinion that, with our current knowledge of the organic, we are not nearly in a position to provide any demonstration of the correctness of any kind of thought on this question.

If it was meritorious for John Tyndall, William Preyer, and Eduard Pflüger to point out the resemblance of the combustion process to the life process itself—fire as the oldest and most widely used parable of life—we cannot express the slightest suspicion based on factual observations that the life process was derived from fire. We know far too little about the achievements of atoms for themselves, and of organic structures, to be able to judge whether a direct transition from fire to life was possible. Likewise, it seems to me superfluous to search the whole universe theoretically for a possible place of origin because we lack any conception of the necessary qualities of this place. For the time being, I think, we can just as easily be satisfied with the hypothesis that the life process began at some stage in the history of the earth; but one does not need to demand of it, as always happens, orderly contractility and consumption, with appropriate regulation of assimilation.

Rather, we only need to imagine that life began simply as a mere assimilation process similar to fire. Gradually, with the appearance and disappearance of innumerable varieties, with a continual increase in lasting properties, quantitative and qualitative "self-regulation" developed in assimilation and consumption. This was probably followed by the origin of reaction qualities already shown in the monerans to an extraordinarily high level in one direction; little by little, perhaps over millions of years, reflex movement was gradually cultivated in a lower form. The further development of reactions such as firmly ordered movement and specific sensory perception undoubtedly followed much later. These are already so much higher than our imagination that no one demands them from the lowest level of life. But the much more difficult acquisition of the properties required before the appearance of these higher reactions is said to have taken place all at once, as a game of chance.

One does not usually consider what it takes to form and move a sham-foot (pseudopodium): how many millions of molecules have to arrange themselves and come closer to one another when stretching out in the form of a ring, and then must do the same afterward when retracting the foot lengthways, and what it takes to acquire these skills. [According to excellent recent work of Gottfried Berthold, Otto Bütschli, Max Verworn, and others, this proposition, of the great complexity of simple organic mass movement, is an inaccurate one, and its difficulty of acquisition has been greatly reduced.]

Reflex movements were probably followed by the development of firmly ingrained, bequeathable tendencies, both in movement and formation, and with them the great principle of formation from processes underlying material exchange, the basic principle of organic morphology. This does not seem to me to be any easier to understand than sensibility; indeed it is rather more difficult, despite the often cited analogy to the formation of crystals, because the latter does not take place from processes with material exchange.

Just as one previously wanted the homunculus to emerge ready-made from the retort, one now demands the same of monerans. This seems to me as if one expected a hurricane to blow tunefully by chance a self-organized work of art such as a Beethoven symphony, or demanded that when old rocks collapse into rubble they should assemble a stylish Doric temple, or that a Papuan might happen by chance to discover integral calculus. If something, whose creation requires the selection of the best for thousands of years, can emerge suddenly in complete form from the hand of chance, why should it not also happen in these cases? Yet these cases are perhaps simpler than the arrangements of particles in the movement of monerans, which is not fixed once and forever but constantly changing.

The developmental steps, from a simple process of assimilation to one with sensibility, from the latter to the emergence of certain tendencies that can be transferred through bequeathal, and from hence to man, do not seem so unequal to me. According to our present-day (admittedly completely inadequate) conception, the principal achievements are perhaps co-equal; but a fourth step will have to be inserted into the total effect that begins with the origin of consciousness, with the gathering together of every experience of the individual into a single total effect. But if the essence of consciousness had already been better analyzed analytically, it would perhaps not seem so essential to us to represent a special step for this kind of abstraction from which the entire rest of the mental pageant may be derived. In any case, however, it seems arbitrary to assume that consciousness is a general property of matter for the sole purpose of saving us from the inconvenience of having to introduce it as newly created in the organic. An endless number of completely novel qualities have appeared in the course of organismal evolution and have been added to the few original ones. We are just as unable to derive their specific quality from the properties of atoms (the material substrate to which they are bound and the functions of which we rightly consider them) as we are to derive consciousness from the properties of the ganglionic cells of the cerebral cortex.

But it is a tendency of our time, which has emerged from striving to reduce the manifold to the simple, to deny these novel qualities and to say that because monerans have the same major functions as higher organisms (nourishment, reproduction, reflex movement) no novel qualities have appeared. For the new are merely descendants, gradual differentiations of the simpler. But is that really why their differentia are less novel? Every chemical change in an organism is a new quality, no matter how "gradually" it emerged from another. Indeed, every transitional stage to a so-called new chemical quality is already a new quality. So long as the elements have not yet been derived from a single element, the chemical qualities are qualities in the full sense of the word. But this remains true, even if (as according to Leucippus, Democritus, and Robert Boyle) all differences were to be derived from quantitative differences, from unequal groupings of molecules of a single basic substance. For different chemical compounds, different groupings of the same elements have different properties and behave differently in actuality. There is therefore no fundamental reason to refuse the assumption that, just as some chemical compounds produce light or electricity, certain chemical processes should not also have the ability to fix the physical-chemical experiences of the individual in the brain, and to the same extent have the ability to excite them again in greater or lesser intensity by the action of stimuli [furthermore to form an "abstraction" of the commonality of several excitations; and finally even an "abstraction" of the commonality of all excitations of the same living being, that is to turn all excitations of this living being into a special excitation. This "*personal abstraction*" then represents *consciousness* or, more correctly, *self-consciousness*].⁹

Our concern here is merely to indicate that perhaps the psychic functions are not something absolutely different from everything else that happens, that like everything else the psychic functions could be derived from one of the many different qualities that are present in organisms and that do not cease to work even if they are denied for a few decades. And their origin will also have been very gradual. It may have taken millions of years before the first abstraction was formed from the most common and sufficiently varied things as a still unconscious excitation of their commonality; and a similar time may have passed before the regular return of a pain after a stroke was understood as not just as regular succession, but rather as more closely related to one another, although this determination of causal relations in particular seems to me to be a relatively easy acquisition. Some older dogs no longer bite the stick with which they are beaten, but bite the legs of the person who beats them.

---

9 In the place of the passage in square brackets, the first edition has {and to let them unite to form a general impression. Whether this self-awareness was the first abstraction, or whether it first arose through other, more frequently occurring, mutually similar impressions than the natural, stronger excitement of what they had in common, we, as nonexperts, cannot discuss}. —Trans.

But it is pure arbitrariness to ascribe every property that has gradually developed (of which one cannot with certainty indicate the first beginning or first appearance) to the lowest organisms or even to inorganic processes. It seems to me that here we have the unsolved problem of the bald head that still perplexes the imagination.

One can ascribe consciousness to monerans just as much as one can ascribe baldness to the bearer of a splendid head of hair. As is well-known, one cannot designate when baldness begins so long as the hairs are pulled out individually. One could just as well claim that one packet of vermillion, among other packets of vermillion, to which aniline blue is added drop by drop, starting below the limit of perceptibility, was from the very beginning a packet of blue because the boundary of the first appearance of blue cannot be certified.

Thus, it seems to me superfluous to say anything further about our present inability to judge the time and place of the first[10] origin of life and the successive appearance of its most important properties. I content myself with having raised my voice for the recognition of new essential properties of the organic, supplementing assimilation, self-movement, and self-division, namely *"overcompensation in replacement"* and *"self-regulation"* [as well as for the *emergence of the first life*—that is to say, of the emergence of the complex of these properties through *"successive cultivation" of the individual basic properties* and of each of them to ever higher degrees of achievement up to perfection].

[Self-regulation was already discussed in first and fourth chapters as the self-organization of the purposive in relation to "morphological functional adaptation" as the shaping of forms. Since we have thus found that self-regulation is an indispensable fundamental property of living things, our earlier assumption of the trophic effect of function, or of functional stimuli, receives further support, because this mode of action represents a most elementary and useful mechanism of "morphological self-regulation." In this, the chapter, which is now concluded, may find its justification although it seems somewhat different in content from the major themes of the book.]

---

10. *Vormaligen* (former) in first edition; *ersten* (first) in second edition. —Trans.

CHAPTER VI

Summary

We saw in the first chapter that Darwin and Wallace provided no account of morphological functional adaptation that can explain the direct self-organization of the purposive even under completely new circumstances. We also saw that such an explanation of direct organization of the purposive was much needed and provides the most dangerous competition for their main principle (the emergence of the purposive—that is to say, that which is lasting—through selection from free variations) and gives it the semblance of a metaphysical principle.

Functional adaptation's effects are manifested in the direct purposive remodeling of organs in an individual's life when they are lastingly changed in the nature and extent of their use by a newly appearing embryonic (or acquired pathological) variation of a part; or when this altered use is enforced by an alteration of external living conditions; or, in human beings, by free will. A new group of effects was added to this long-known mode of action, consisting, on the one hand, in the formation of the static structure of bones and connective tissues as well as in the formation of the corresponding dynamic structure of organs formed from smooth muscle fibers, and second in the perfect adaptation of vascular walls to the form of the bloodstream. This second group of subtle adaptations made the problem of explaining morphological functional adaptation much more difficult,

230 THE STRUGGLE OF PARTS

whereas the previously known coarser remodeling had been supposed to be derivable merely from functional changes of the blood supply.

The second chapter showed that not everything that happens in organisms is strictly determined, molecule by molecule, down to the smallest detail; how, as a consequence of material exchange and change in external conditions, this is also impossible; and that instead, from continual variations in the qualities of the parts a struggle occurs between the new and old qualities over food and space (as well as a self-eradication of non-lasting qualities); and that this must have occurred in organisms from time immemorial.

As we have seen, in this process of selection, called *a potiori* the "struggle of parts" or "selection of parts," only the qualities that are most vigorous under existing conditions triumph and remain alone on the field. In those organs often acted upon by stimuli—for example, that trigger function—the victorious properties are those whose ability to assimilate is most strengthened by action of the stimulus.

Processes are thus cultivated by the selection of parts that are capable of producing the phenomena of functional adaptation. And this is a consequence of the struggle of "similar" structures with one another: namely, the struggle among living cell parts [isoplassons, autokineonts, automerizonts, idioplassons] and the struggle among cells of the same tissue. Among other things, the ability of individual organisms to become accustomed to harmful constituents of food, such as organic and inorganic poisons, is explained by the self-eradication of those parts that are not lastingly able to survive under these particular dietary conditions. Perhaps immunity after vaccination or survival of illness also partly rests on the same principle.

On the other hand, the struggle between "dissimilar" tissues and organs leads to the greatest possible utilization of space within the organism and to the development of an inner harmony, a morphological balance corresponding to the physiological significance of the parts for the whole.

The importance of Darwin's principle of the struggle of individuals for the origin of variety and for adaptation to external conditions is not in the least restricted by these outstanding achievements of the struggle of parts. Rather, the relationship between the two kinds of struggle is such that the struggle of parts selects and cultivates those special substances [or more correctly instantiations of processes] that are also suitable in the struggle of individuals, in general those that are most vigorous and most strongly reacting.

Thus, while the struggle of parts brings about so-called purposiveness [*Zweckmässigkeit*] within organisms, better called lastingness [*Dauerfähigkeit*], and the highest performance [*Leistungsfähigkeit*] of the same in the general dynamic sense, the simultaneous struggle for existence among individuals brings about an outward lastingness, proving itself in the individuals' external conditions of existence.

However, for this struggle of parts to have been effective, properties were assumed to have occasionally appeared in organisms that won victory and expansion to the point of solitary existence. The availability of such occasionally appearing properties first had to be demonstrated. Evidence for this was advanced in the third chapter, I believe in a manner sufficient for the first foundation of the whole. This chapter was about the property of the protoplasm of the various tissues to be excited by the functional stimulus, not only to *specific activity* but also *assimilation* (whether directly or indirectly) for the formation of new corresponding substance, as well as for replacement and overcompensation of what has been consumed. This quality encompasses the principle of the functional self-organization of the purposive or lasting.

The production of static structures in bone tissue that are adapted to new conditions of pressure and tension speaks in the most obvious way for this cellular property. Moreover, the rapidly progressing degeneration of actively functioning parts when the functional stimulus is entirely excluded seems to prove the same thing for muscles, nerves, and glands. I also showed that the previous explanation of activity hypertrophy and inactivity atrophy by alterations of blood supply to the organs connected with function is completely inaccurate and is contradicted by the most general biological empirical knowledge. The impossibility of such causation was directly proven by particular facts.

After the trophic effect of functional stimuli in the tissues was thus established as far as was possible, the special morphological mode of action of this principle was discussed in particular. The fourth chapter demonstrated that this property is indeed able to directly produce the purposive everywhere, both quantitatively and formatively. Through the ability of the struggle of parts to cultivate such qualities, a much higher inner perfection (a purposiveness of the functioning parts down to the last living particle) could and should have arisen much more rapidly than if it had been created and developed by selection of formal variants in the struggle for existence among individuals (as according to Darwin's theory).

[With sufficient centralization of functional stimuli in an individual's cerebral center, these formative functional correlations provide a principle of increased influence of the organs on one another that is purposive for (increases the lastingness of) the whole individual, that sufficiently and mechanically explains the individual's direct adaptation to new conditions, and at the same time really creates teleological things insofar as it is moved to activity by the will.

In the earlier stages of individual development, the parts exhibit self-sufficient abilities to grow and shape themselves without the need for functional stimuli. Therefore, the life of each part can be divided into two phases: a so-called embryonic period independent of function (thus, in this sense, self-sufficient life) that gradually merges into the later period of "functional stimulus-life."]

Finally we looked at the organic in general and tried to approach its "essence." We recognized that its most necessary property is lastingness under "changeable" external conditions. From this it followed that a fundamental property of the organic was the ability to "self-organize" the necessities for self-preservation under conditions of "change." This is the morphological functional adaptation beginning with assimilation as a special property and continued through many morphological "self-regulatory mechanisms," as well as the transitory "pure functional adaptation" to current needs. The latter two are both types of "*self-regulation.*" Overcompensation in replacing what has been consumed mediates the first, morphological type of self-regulation. As simple "growth," this overcompensation is necessary for the proliferation of organisms. Thus, "*self-regulation*" and "*overcompensation when replacing what was consumed*" are, next to *assimilation, the most essential "general" properties of all organic activity.* Reflex motions and self-division rank only after these in importance, even if they are common to all organisms and special achievements of them.

At the same time was revealed a hitherto nowhere discussed possibility of the origin of the first, lowest life from suitable accidental variations of earthly events over the course of very long periods of time, by the *"successive" self-cultivation of the "lowest" process properties of life* following one after another (beginning with assimilation, and then overcompensation in assimilation, expanding to reflex movement itself), whereby each individual property could gradually cultivate itself to ever greater perfection. Furthermore, with the cultivation of stimulating substances, a hint of a possibility of the origin of abstraction and consciousness was shown.

THE ABOVE TREATISE has perhaps contributed something to complete and round off the "general theory of development" of organisms by showing which general properties alone could win duration in the interaction of natural events and which must have been raised from one level to the next through summation or, more accurately, through their variations surpassing each other. Although the causes of the existence, of the preservational persistence, of these properties may have been demonstrated, this has of course not furthered in the least our knowledge of the causes of their origin and coming into being. This is the problem of the relevant molecular events by which certain causes took place in a certain way according to physicochemical laws.

But to demand such a thing from mere principles of maintenance and increase, which form this general theory of the development of organisms, would be the same as asking a mathematician to determine the rate of heat vibrations purely theoretically—that is, to determine a concrete happening that occurs through quantities purely from qualities (about which we do not know enough). Some think this is possible. After I had described the methods for determining the airspeed of cannonballs, a high-school professor and excellent philologist once asked me, with astonishment, why such cumbersome empirical methods were required for what could be easily calculated. Not a few naturalists implicitly endorse his view when they reproach the doctrine of descent for being unable to explain a single physiological process.

With these direct developmental events, the basic morphological problems remain unexplained: the formation of direction and shape from chemical processes that are inherently directionless and shapeless, and the embryonic formation of the complex from the simple without any external differentiating cause. As before, we stand before these everyday phenomena as incredible, incomprehensible wonders.

The Afterlife of
Der Kampf der Theile im Organismus

David Haig

Early Reactions

The best available guide to the early reception of *Der Kampf der Theile* is Roux's foreword to its 1895 second edition (included in this translation) in which Roux scrupulously collected notices of his book, both good and bad. Roux applauded the perspicacity of the positive reviews but accused the negative reviewers of reading his book cursorily. He singled out George Romanes's review (1881a) in *Nature* as "probably the reason it has remained virtually unknown in England." Roux, like most authors, had high hopes for his beloved child and was disappointed that others had not shared his enthusiasm.

The foreword and footnotes to the second edition provide evidence of a priority dispute between Roux and William Preyer who had taught him at Jena. Preyer delivered a lecture at Prague on December 3, 1878 (the date is provided in the 1880 printing), entitled "Die Konkurrenz in der Natur" (Competition in nature) that included a discussion of competition within organisms. Preyer's lecture was printed three times in the ensuing three years: first in the February 1879 issue of *Nord und Süd* published by Schottlaender in Breslau (1879); second in a collection of Preyer's popular lectures published by Pætel in Berlin (1880); third as a thirty page "separate" published by Schottlaender in Breslau (1882).

Roux submitted his doctoral thesis in Jena in 1878, before Preyer's lecture, worked for a year in Leipzig, then moved to Breslau in October 1879. *Der Kamp der Theile* was published by Engelmann in Leipzig in February 1881. At some point Preyer became aware of its contents, which he believed had slighted his own presentation of similar ideas in "Die Konkurrenz in der Natur." Preyer stayed with George Romanes in the summer of 1881. At that time, Romanes reported to Darwin that Preyer knew Roux but did not think much of his book. It is possible that Preyer first learned of *Der Kampf der Theile* from the copy Roux had sent to Darwin that Darwin forwarded to Romanes (see the introductory essay of this translation for the chronology of events associated with Romanes's review of *Der Kamp der Theile*). In this scenario, Preyer arranged for a new printing of his lecture in Breslau (1882), where Roux was now Privatdozent, to emphasize his priority in print.

In the 1895 edition of *Der Kampf der Theile*, Roux quoted from Preyer's essay, which he says he first knew from the "separate" of 1882. Roux emphasized that he had written a paper (1879) that discussed the struggle of parts before he had read Preyer's essay. Moreover, Roux wrote, neither Preyer (1879) nor himself had acknowledged Ernst Haeckel's (1876) earlier discussion of a struggle of parts. Roux thus undercut Preyer's claim to priority in print. Roux acknowledged that he had attended lectures by Preyer and Haeckel when he was a student at Jena in the early 1870s and that these lectures perhaps might have planted the germ of the idea in his mind. Roux further emphasized that Preyer discussed the harmony that arises from competition among dissimilar parts whereas *Der Kampf der Theile* was principally concerned with competition among similar parts.

August Weismann was an early reader of *Der Kampf der Theile*. Weismann's inaugural address as pro-rector of the University of Freiburg was delivered on June 21, 1883, and published that August as *Über Vererbung* (in English as "On Heredity," 1891). In his address, Weismann argued for the continuity of reproductive cells across generations and categorically rejected the inheritance of acquired characteristics. He accepted that a struggle of parts took place during development but rejected the idea that the effects of this struggle could be inherited (88–89). This was a major challenge to one of the central tenets of *Der Kampf der Theile*. Roux's revisions for the second edition of *Der Kampf der Theile* respond to that challenge with limited success.

Weismann's *Das Keimplasma* (1892) contained a brief mention of Roux's struggle of parts. Weismann acknowledged the body's ability to respond adaptively to abnormal conditions but thought it would be a great mistake to relate most of normal ontogeny to the struggle of parts (143). He also considered the possibility of a struggle of parts among biophores (349) and determinants (365). *The Germ-Plasm* (1893), an English translation of *Das Keimplasma*, appeared the following year.

On May 2, 1894, Weismann delivered the Romanes lecture at the Sheldonian Theatre in Oxford with George Romanes in attendance (Romanes would die on May 23). In this lecture, Weismann discussed *Der Kampf der Theile* at length. Variability, heredity, and a struggle for existence were present within bodies, he argued. Therefore, intra-individual selection must inevitably take place. "It is not the particular adaptive structures themselves that are transmitted, but only the quality of the material from which intra-selection forms these structures anew in every individual life. . . . The great significance of intra-selection appears to me not to depend on its producing structures that are directly transmissible,—it cannot do that,—but rather consists in its causing a development of the germ-structure, acquired by the selection of individuals, which will be suitable to varying conditions. *Intra-selection effects the special adaptation of the tissues to special conditions of development in each individual*" (1894, 15–16, emphasis in original).

In a posthumously published volume, Romanes invoked Roux's struggle of parts to explain the degeneration of organs that had fallen into disuse (1895, 298f). Six years previously, Romanes (1888) had labeled Weismann and his followers "neo-Darwinian" because the true followers of Darwin believed that the effects of use and disuse could be inherited. Romanes had earlier written an excoriating review (1881c) of Samuel Butler's *Unconscious Memory* (1880), in which Butler used neo-Darwinism to refer to the doctrine of Charles Darwin as distinguished from the "original" Darwinism of Erasmus Darwin. Therefore, Romanes (1888) may have subconsciously borrowed Butler's term as a label for Weismann and his followers.

Weismann's Romanes lecture elicited a broadside from Herbert Spencer (1894) in his long-running dispute with Weismann over the inheritance of acquired characteristics and the adequacy (Weismann) or inadequacy (Spencer) of natural selection to explain adaptation. In a footnote to his polemic, Spencer wrote that "Professor Weismann is unaware that the view ascribed to Roux, writing in 1881, is of far earlier date." Spencer then quoted passages from his own "The Social Organism" (1860) and *Principles of Sociology* (1876) in which he had discussed competition for nutrients between different parts of an organism. "Though I did not use the imposing phrase 'intra-individual-selection,' the process described is the same" (1894, 587).

Weismann (1895) responded to Spencer's claim to have described the struggle of parts twenty years before Roux with a footnote of his own. The passages Spencer had quoted presented an analogy between processes within organisms and within societies, but Spencer had never worked out how such processes could give rise to purposive histological structure. "There is still a long way to go from the one-time flash of an idea to its implementation. . . . In this case one may well say that the entire scientific world would hardly have welcomed the book in question by Wilhelm Roux (1881) as something completely new and as a significant advance in knowledge" if it had simply repeated ideas already formulated by Spencer.

In his response to Spencer, Weismann wrote that Roux let "this process of intra-selection take place not only between tissues and tissues, between cells and cells, but also between the components of the cells. If, as there is no doubt about it, it exists at all, it must take place between the life-units of every degree, and therefore also between the determinants of the germ-plasm" (1895, 11). The next year, Weismann developed this idea at length. The struggle of parts within the germline could result in heritable changes. He called this "germinal selection" as distinct from somatic "intra-selection" whose effects were not inherited (1896).

Weismann's lectures at the University of Freiburg were published as *Vorträge über Descendenztheorie* (Weismann 1902, 1904). Lecture XII addresses intra-selection as a preamble to Weismann's rejection of the inheritance of acquired characters. In Weismann's view, Roux's "functional adaptation" was essential for normal development but had evolved by *personal* selection: "every new kind of glandular, muscular, or nerve cell, such as have arisen a thousandfold in the course of phylogeny, can only have resulted from a struggle between individuals which turned on the possession of the best cells of a particular kind, not from a struggle between the cells themselves, since these would gain no advantage from serving the organism, as a whole, better than others of their kind" (1904, 250). The same three factors that are active in personal selection—variation, inheritance, and competitive struggle—are present during histonal differentiation. Therefore, histonal selection or intra-selection must inevitably occur during somatic development. Roux's "functional adaptation," whereby different cells and tissues respond to functional stimuli, is responsible for much of the adaptive form of the organism, but its effects are not inherited.

Weismann's Lectures XXV and XXVI address germinal selection. By contrast to the functional adaptation achieved by intraselection of somatic cells, which was nonheritable and must be regenerated anew each generation, Weismann believed that germinal selection could contribute to the disappearance of functionless or useless organs. Once a trait was no longer subject to positive personal selection (an absence Weismann called "panmixia"), germinal selection would contribute to the elimination of a trait no longer under personal selection.

Elié Metchnikoff's "*La lutte pour l'existence entre les diverses parties de l'organisme*" (The struggle for existence among the various parts of the organism) (1892) used as its title a close paraphrase of *Der Kampf der Theile im Organismus*. Metchnikoff commended the speculations of the very talented Monsieur Roux. In Metchnikoff's view, the internal struggle consisted primarily of differential resistance of cells to the devouring activity of phagocytes. The destruction of weaker cells by phagocytes strengthened the organism and was central to the developmental creation of form.

Yves Delage discussed Roux's struggle of parts—Delage called it *sélection organique*—at length (1895, 394–395, 724–742). Delage found much of value in Roux's hypothesis, especially in the role of functional stimuli in shaping organismal form, but Roux left questions of heredity unresolved, in particular the inheritance of acquired characteristics and how characters were represented in the germ-plasm. Ludwig Plate's *Selectionsprinzip und Probleme der Artbildung* (1899, 1913) also discussed Roux's struggle of parts at length under the heading of *Intralkampf*. Although Plate considered Roux's book to be "epoch-making" (1913, 337), he found little need for a struggle of parts during development. In his view, embryonic development was controlled by heredity, and the personal struggle for existence was entirely adequate to explain adaptation. Vernon Lyman Kellogg largely followed Plate's critique but saw in Roux's theory a concession that selection alone was inadequate to explain fine adaptation (1907, 201–208). In his refutation of Darwinism, the prominent cell biologist Oscar Hertwig (1922) rejected the struggle of parts in favor of a harmonious system of coordinated development.

Other contemporary reactions to *Der Kampf der Theile* included Johannes Unbehaun's *Versuch einer philosophischen Selektionstheorie* (1896), in which Unbehaun presented a generalized theory of selection in the organic and inorganic world complete with extensive differential equations. August Pauly's *Darwinismus und Lamarckismus* (1905) found much that was commendable in the struggle of parts but was unable to accept Roux's derivation of the purposiveness of living things from mechanical principles. He considered this "one of the greatest psychological oddities in the history of Darwinism" (84). Pauly instead ascribed organic purposiveness to his own peculiar brand of psychophysical teleology.

Der Kampf der Theile figured prominently in *Form and Function* (1916), Edward Stuart Russell's account of the beginnings of causal morphology. Russell ascribed Roux a seminal role in the unification of form and function, with function ultimately determining form (313–335). Russell emphasized the Lamarckian elements in Roux's theory including the inheritance of acquired characters. Jan Boeke (1921) rejected Roux's struggle of parts because it did not account for the harmony of the organism.

A Brief Bibliometric Comparison

Roux's *Der Kampf der Theile* (1881) and Weismann's *Das Keimplasma* (1892) both presented hypotheses that their authors intended to significantly extend Darwinism. Weismann's book soon appeared in English translation as *The Germ-Plasm* (1893). A simple citation count using Google Scholar (May 2023) found that *Das Keimplasma,* with 410 citations, and *The Germ-Plasm,* with 700 citations, together had received more than twice as many citations as the 510 citations to Roux's book. This

Table 1. Number of articles citing Roux's and Weissman's books by language

	Der Kampf der Theile	Das Keimplasma	The Germ-Plasm
German	46	22	0
English	12	2	29
Italian	2	1	0
French	1	1	0
Russian	1	0	0

conformed to my expectation that Weismann's book was better known, primarily in English translation.

What surprised me when I looked at historical trends was that *Der Kampf der Theile* received sixty-two citations before 1930, about the same number as the fifty-five citations to the German and English versions of *Das Keimplasma*. My expectation had been that Weismann's book would have been better known in this period as well. When I broke down the citations prior to 1930 by language of citing work, a clear pattern emerged. Roux's book received more than twice as many German citations as Weismann's books, but Weismann's books received more than twice as many English citations (and these mostly to the English translation). My expectation had been biased by my better knowledge of English language publications from this period.

Subsequent Responses

This section presents an overview of reactions to *Der Kampf der Theile* after Roux's death in 1924. Most of the twentieth century was a fallow period for biological interest in *Der Kampf der Theile,* partly because the book advocated what was seen as a false theory of Lamarckian inheritance, partly because of the developing disciplinary boundaries between developmental and evolutionary biology, partly because the book did not exist in English translation.

Bartlomiej Swiatczak (2023) has published an impressive history of ideas about somatic selection from the eighteenth century to the present day. He found numerous expressions of the idea of internal selective processes during the nineteenth century, both before and contemporaneous with Roux, but found these ideas largely disappeared around 1900. Swiatczak ascribes this change to the "eclipse of Darwinism" at this time and the onset of a holistic orientation toward biology in which the whole determined the parts.

During the first two-thirds of the century, organisms were usually thought of as harmonious wholes whose parts were adapted by natural selection to serve a common good. Sometimes this was seen as the good of the organism, but often it was seen as the good of the group or the species. This perspective began to change in the 1960s with vehement rejections, and equally vehement defenses, of group-level adaptations. On one side, the beneficiaries of adaptations were seen as restricted to individual organisms, even individual genes. On the other side, advocates of group-level adaptation promoted theories of multiple levels of selection in which adaptations could exist both below and above the level of the individual. These debates have been associated with a gradual revival of interest in *Der Kampf der Theile* within evolutionary biology. In more recent years, there has also been a revival of interest within developmental biology, in part associated with attempts to reintegrate evolutionary and developmental biology (so-called evo-devo).

Roux's aim in *Der Kampf der Theile* was to provide a mechanistic account of the exquisite functionality of bodily parts in their service to the whole. In his dynamic model of development, the strength and vigor of parts were enhanced by their use (the stimulus of function was present) or diminished by their disuse (the stimulus of function was absent). Roux further believed that the external Darwinian struggle for existence among individuals was inadequate to explain the internal purposiveness of living things which he thought could be accounted for by Lamarckian inheritance of the effects of use and disuse combined with an internal struggle in which the differential survival and proliferation of organismal parts played a key role in the fine-tuning of efficacious structures.

At the time of writing the first edition of *Der Kampf der Theile,* Roux viewed inheritance as akin to a form of memory in which the structure of ancestral parts was remembered by descendant parts. Therefore, ontogeny was intimately linked to phylogeny. Morphological changes that occurred during an individual's somatic development would necessarily be bequeathed to its offspring and inherited by their descendants. This view of inheritance was challenged by Weismann's hypothesis of the continuity of the germ-plasm (1883, 1891, 1892) and then by the development of modern genetics after 1900. Form came to be viewed as determined by the inheritance of determinants of form, not by memories of earlier forms.

The distinction between forms and inherited determinants of form (we would now call them genes) has important consequences for how competitive processes during development are conceptualized. As Weismann (1894) recognized, mechanisms of functional adaptation could have evolved by selection among individuals without requiring the direct bequeathal of somatically acquired functionality. Questions about the importance of functional adaptation during development could be separated from the question whether or not the outcomes of these processes were directly inherited. In particular, inherited determinants could generate competitive processes during ontogenetic development that achieved an optimal proportion of parts in the most appropriate locations. Such processes could have evolved by competition between alternative determinants of form (we would now call them alleles) by processes summarized as "natural selection."

Roux proposed that functional adaptation enabled organisms to automatically adjust their fine structure to new conditions and major changes of form. This facilitated evolutionary change because multiple characters could exhibit a coordinated response to an environmental change rather than each character needing to evolve an independent response. Blood vessels would be delivered where they were needed. Bones would adjust to new forces to which they were subjected. Once functional adaptation is separated from the Lamarckian hypothesis that developmental changes can be directly inherited, Roux's ideas harmonize with modern ideas about the evolutionary importance of "developmental plasticity" (West-Eberhard 2003) and "facilitated variation" (Gerhart and Kirschner 2009). The organism is able to respond adaptively to changed conditions, and this adaptive plasticity facilitates subsequent evolutionary change.

Modern models of cellular competition come in one of two flavors. Models of the first flavor invoke cellular competition as a means of achieving functionality within the body and emphasize the importance of stochastic processes in determining cellular fate. Models of the second flavor invoke a conflict between levels of selection in which cellular selection undermines bodily cohesion and organismal fitness. In models of the first flavor, it is the organism that benefits whereas in models of the second flavor it is the determinants of form resident in successful cell lineages that benefit, sometimes at the cost of the organism.

For the most part, Roux assumed that the properties favored by the struggle of parts within organisms would be properties that also benefited individuals in their own struggle for existence. Therefore, his ideas about cellular selection are mainly of the first flavor. However, Roux occasionally discussed the possibility that internal struggles could have maladaptive outcomes for individuals. In those cases, Roux reasoned that these individuals would be eliminated by individual-level selection. Therefore, the long-term effect of the internal struggle of parts would be to enhance organismal adaptedness in the external Darwinian struggle among individuals. These discussions have a hint of the second flavor.

Cellular Selection as a Cause of Functional Adaptation

My discussion of the biological reception of Roux's ideas will proceed under three headings. In this section, I will discuss models of cellular selection of the "first flavor" in which stochastic variation is generated within bodies during development and functional adaptation is achieved by selection from among the variants so generated. In the next section, I will briefly consider models of cellular selection of the "second flavor" in which selective processes acting on stochastic variation within the body result in organismal maladaptation. In the following section, I will discuss the special case of orthopedics and bone biology. Modern interest focuses on the importance that Roux attached to physical forces in the remodeling of bone.

In cellular-selection models of the first flavor, competition among cells achieves adaptive outcomes because competitive *mechanisms* have evolved by Darwinian selection to enhance individual fitness. James Michaelson (1987) contrasted instructional and selectional models of the determination of cellular fate. He cites Roux (1881) as his earliest example of a selectional model. Under instructional models, cells are told what to do, and they do what they are told. Under selectional models, cells that find themselves in the right place, or perchance do the right thing, survive whereas those that are in the wrong place, or perchance do the wrong thing, are eliminated. Selectional models invoke the generation of random variability among cells to provide the raw materials for cellular selection.

Selectional theories were developed during the twentieth century to explain the functionality of the adaptive immune system (Jerne 1955; Burnet 1959) and the nervous system (Changeux, Courrège, and Danchin 1973; Edelman 1987). The cellular basis of immunity was poorly understood in 1881, and Roux had little to say on the subject. Metchnikoff (1892) invoked the struggle of parts in his discussion of phagocytes, but I have found no evidence of a significant influence of *Der Kampf der Theile* on selectional theories in immunology. Leon Chernyak and Alfred Tauber, in their encomium for Metchnikoff, considered Roux's "orthodox Darwinism" metaphysically trivial compared with Metchnikoff's account of phagocytosis with its "significant modification and rethinking of Darwinism" (1990, 238f).

In contrast to an apparent lack of influence in immunology, *Der Kampf der Theile* was recognized during the twentieth century as a precursor of selectional models in the nervous system. One reason to recognize a forgotten precursor is to undercut the presumptions of a present rival. Roux's ideas on the struggle of parts are given progressively greater prominence in successive editions of Marcus Jacobson's *Developmental Neurobiology*. In the first edition (Jacobson 1970, 160–161), *Der Kampf der Theile* is said to have influenced Ramón y Cajal's ideas of competitive struggle for space and nutrition among neural outgrowths. In the second edition (Jacobson 1978, 207, 305–306), Roux is acknowledged as the first to recognize the importance of selective processes in the development of the nervous system. Then, in the third edition (Jacobson 1991, 92, 231–232, 314, 416), Roux's theoretical position is presented as the origin of "so-called neural Darwinism." In Jacobson's judgment, the theory of competition and selection had been "reinvented in more modern terms . . . without acknowledging Roux's priority in spite of attention having been drawn to his contribution in both previous editions of this book" (Jacobson 1991, 232). Gerald Edelman's *Neural Darwinism* (1987) is the obvious unnamed target.

Jean-Jacques Kupiec (1986, 1997) developed models of embryogenesis in which Darwinian selection of randomly generated variation played an important role. These ideas were developed independently of any knowledge of Roux's struggle of parts. Thomas Heams, who was then a doctoral student of Kupiec's, was struck by a comment of André Pichot (2003) that Kupiec's pan-Darwinism was simply a restatement of ideas originally presented by Roux in 1881. Heams obtained a copy of the first edition of *Der Kampf der Theile* in about 2008 (personal communication) and presented a paper on Roux at a 2011 conference in Lyon organized by Kupiec on "Chance at the Heart of the Cell." In this talk, Heams presented Roux as a pioneer of the idea that stochastic variability among cells, rather than homogeneity, lay at the heart of developmental processes. The resulting article (Heams 2012) is perhaps the most substantial engagement with Roux's text in any biological context since Roux's death. Heams's article provided the impetus for the French translation of *Der Kampf der Theile* (Roux 2016) for which Heams provided the preface. Michaelson (1987) and Begoña Díaz and Eduardo Moreno (2005) also discuss the role of cell competition in general development and cite *Der Kampf der Theile* as an early selectional model. Kate Vasquez Kuntz and colleagues (2022) describe how Roux's struggle among cells could enhance organismal adaptation.

Although the publications reviewed here focus on competitive processes among cells, *Der Kampf der Theile* invoked multiple levels of selection: among living molecules within cells, among cells within tissues, among organs within individuals, and among individuals. This has been considered a modern feature of *Der Kampf der Theile*. Thus, Leo Buss (1987) presented Roux's book as a forgotten precursor of theories of multilevel selection and Stephen Jay Gould's *Structure of Evolutionary Theory* (2002) contains an extended discussion of *Der Kampf der Theile* (208, 210–214), which he considered one of the "most important untranslated documents of 19th century German evolutionary biology" because of its invocation of a hierarchy of structural levels (208). This evaluation was part of Gould's extended campaign to undercut theories that considered the individual (or gene) as a privileged level of selection. In recent years, Michael Levin and his coworkers have developed models of morphogenesis in which competitive processes occur at multiple levels, with subagents at each level competing to achieve level-specific competencies (Gawne, McKenna, and Levin 2020; Levin 2021). Competencies of lower levels are a necessary substrate for the development of competencies at higher levels. Levin cites Roux's "struggle of parts" as a precursor of these ideas.

Cellular Selection as a Cause of Organismal Maladaptation

Theories of multilevel or hierarchical selection rose to prominence in evolutionary theorizing during the 1970s and 1980s associated with debates about what entities should be considered the beneficiaries of adaptations and whether cooperative behaviors could emerge at higher levels of selection despite selection at lower levels favoring selfishness. In these models, heritable genetic variation and selection occur at each level of the hierarchy, but the adaptations achieved at one level may conflict with what is adaptive at another level. In cellular selection models of the second flavor, what promotes cellular fitness in the competition among cells within bodies may undermine the organismal fitness of the body composed of those cells.

Some have interpreted Roux's struggle among cells to be a precursor of evolutionary theories of cancer (Yanai and Lercher 2020), but this misunderstands Roux's intent. In one passage, Roux writes, "A discussion of tumors does not really belong here because we would be dealing with abnormal formations and abnormal stimuli. However, we do not want to omit a cursory look at the cause of their origin, in order to possibly obtain a useful analogy to the trophic effect of the surrounding functional stimuli." For Roux, the struggle of parts enhanced organismal function, and tumors were of interest for the clues they provided about normal processes. Thus, Roux discusses how metastases of a cancer induce the formation of nourishing capillaries that promote the growth of the tumor, but his focus is on what these observations say about the ability of blood vessels to adapt to the nutritional needs of other tissues. For Roux, tumors of unlimited growth were mostly "surplus remnants of embryonic tissue" that were able to grow in the absence of functional stimuli; they were thus peripheral to his interest in explaining the well-ordered development of the body. Perhaps for this reason, Michel Morange (2012) concluded that Roux did not present a well-developed evolutionary theory of cancer.

Bone Remodeling and Wolff's Law

Most biological citations since 1980 of *Der Kampf der Theile* occur in works on bone remodeling. Roux's ideas about functional adaptation, rather than his ideas about the struggle of parts, have dominated these discussions. The idea that the structure of bone is remodeled by the physical forces to which bone is subjected (consider the thicker bones in the racquet arm of a professional tennis player) has been commonly known as Wolff's law since the publication of Julius Wolff's *Das Gesetz der Transformation der Knochen* (The law of the transformation of bones) in 1892. The recent surge of orthopedic citations of *Der Kampf der Theile* reflects a reappraisal of the relative contributions of Roux and Wolff to understanding adaptive changes in bone structure with some suggesting that Wolff's law might better be named "Roux's law" (Lee and Taylor 1999; Cowin 2001).

The first edition of *Der Kampf der Theile* (Roux 1881) described the effects of use and disuse on the fine structure of bone to illustrate functional adaptation, approvingly citing earlier work by Julius Wolff. Four years later, Roux (1885) published a detailed analysis of an unusually healed tibial fracture that had been compensated by pronounced thickening of the fibula. Roux wrote that this study had been intended to be included in a book "on the transformation-law of bones" then being prepared by Wolff. Seven years later, Roux's specimen features prominently in Wolff's *Das Gesetz der Transformation der Knochen* (1892), which Roux favorably reviewed (1893). In "Die Lehre von der functionellen Knochengestalt" (The theory of functional bone-shape), Wolff (1899) emphasized that he had been working on the subject since the 1870s, before Roux, but when Wolff's theories came under attack from the mathematician Ferdinand Bähr, Wolff (1899) turned to Roux for mathematical advice and quoted four pages of a letter written to him by Roux. Both Roux and Wolff could be pugnacious defenders of their own claims to priority, but in their public dealings with each other they expressed what appears to be genuine mutual esteem.

The recent reappraisal of Roux's and Wolff's relative contributions to bone biology can be traced to Heinrich Roesler's (1981) and Jos Dibbets's (1991) close reading of the nineteenth-century literature in German. Roesler (1981) concluded that Wolff lacked a deep understanding of the relevant physical processes and believed that bone grew by interstitial deposition of new bone rather than apposition of new bone and resorption of old bone. Roux, without openly disagreeing with Wolff, introduced the idea of a control process whereby bone was remodeled by apposition and resorption in response to local forces. Dibbets (1991) takes the stronger position that Wolff "was wrong on all the biological aspects concerning 'his' law" and that the best parts of *Das Gesetz der Transformation der Knochen* (1892) had been cribbed from Roux. In Dibbets's view, Wolff was confused about bone remodeling but was an effective propagandist in establishing his own priority.

History of Biology

The extensive literature on Roux's contributions to embryology by historians of biology focuses on Roux's work after *Der Kampf der Theile*. Roux is seen as having championed an experimental approach to embryology in contrast to the speculative excesses of Haeckel's evolutionary embryology. In standard historiographies, Roux's treatise is usually passed over rapidly, if mentioned at all, before addressing the rise of *Entwicklungsmechanik* and experimental embryology. For example, Garland Allen (1975, 25–28) saw Roux as "revolutionizing" the field of embryology in particular and biology in general through his advocacy of experimentalism. Allen's account of Roux's career jumps from his training with Haeckel in Jena to the publication of his first set of experiments on frogs in 1888 without mentioning the intervening publication of *Der Kampf der Theile*. Jane Oppenheimer (1967, 66, 159) mentions the book as developing an analogy between the struggle for existence of organisms in nature and competition among the parts of an organism during development. Jane Maienschein's (1994) history of the origins of *Entwicklungsmechanik* mentions Roux's book only briefly: "Whereas Roux had at first followed his teacher Haeckel's emphasis on evolution and competition of hereditary units, in a 'struggle of parts' . . . by 1883 he had moved beyond Haeckel" (48). Lynn Nyhart (1995) identifies *Die Kampf der Theile* as Roux's "first major theoretical contribution" (283) but portrays his theoretical contribution as the application of Darwinian adaptation to developmental processes without mentioning that Roux, unlike Darwin, invoked selective processes within the individual. Victor Hamburger (1997) describes *Der Kampf der Theile* as an early work written before Roux had rebelled against the evolutionary approach to development he had learned from Haeckel and before Roux had become a champion of experimental methods in embryology.

There has been little discussion of the contents of *Der Kampf der Theile* and most of this concerns the question of whether Wilhelm Roux's ideas influenced August Weismann. Reinhard Mocek (1974) discusses the multiple levels of selection invoked by Roux in *Der Kampf der Theile* but considers Roux's "struggle of parts" to be quite different from Weismann's "germinal selection" (42). On the other hand, Frederick Churchill's biography of August Weismann describes Roux's *Der Kampf der Theile* as a "curious" book containing a "flight of speculative fantasy" (2015, 265) but concludes that Weismann's theory of germinal selection was strongly influenced by Roux's struggle of parts (266, 424, 465). Ernst Mayr's *The Growth of Biological Thought* (1982) did not mention *Der Kampf der Theile* but discussed Weismann's germinal selection, whereas Mayr (1982) cited *Der Kampf der Theile* in defense of the idea that the organism is a single interacting system and a compromise between competing demands. Mayr (1992) published some of his student notes from 1926 in which he invoked *Der Kampf der Theile* in support of Lamarckian inheritance.

Richard Weikart's (1993) account of the origins of German Social Darwinism cited *Der Kampf der Theile* as an application of Social Darwinist ideas of individualistic competition within the organism and suggested that William Preyer used Roux's ideas of the struggle of parts to bolster his case for individualist competition in society. This appears to be a misreading of the historical record because Preyer's ideas were published before Roux's. In his study of the metaphor of the body as a society of cells, Andrew Reynolds (2007) argued that *Der Kampf der Theile* was influenced by Rudolf Virchow's and Ernst Haeckel's ideas of the cell-state.

Nietzsche Scholarship

Friedrich Nietzsche found in *Der Kampf der Theile* an affirmation of the importance of conflicts within the organism and a diminution of the importance of the external Darwinian struggle among organisms. He wrote, "The individual itself as a struggle of parts (for nourishment, space, etc.); its development bound to a victory over and complete dominance of individual parts; to a withering, an 'organ-becoming,' of other parts . . . the essence of the life-process is precisely this stupendous formative violent power, hewing form from within, using, exploiting, the 'external circumstances'—. These new forms, forged from within, are not formed for a purpose; but in the struggle of parts a new form will not for long remain unassociated with a partial usefulness and then, in accordance with its use, will develop more and more perfectly" (Nietzsche 1906, §647; 2017, 367). An influence of *Der Kampf der Theile* on this passage would be obvious to anyone who had worked their way through Roux's text.

Charles Andler (1928, 1958) appears to have been the first to identify an influence of *Der Kampf der Theile* on Nietzsche's evolutionary thought. In Andler's view, Roux's hypothesis contributed to Nietzsche's idea that an internal struggle of parts strengthened and consolidated the organism. Wolfgang Müller-Lauter (1978, 1999) presented a detailed analysis of Nietzsche's creative reinterpretation of Roux. In Müller-Lauter's interpretation, Nietzsche gained the ideas of a struggle of parts inaccessible to consciousness and an absence of complete centralization within the body from Roux. Nietzsche, however, gave greater prominence to will and command than Roux, who emphasized purely mechanical processes. Others who have explored the influence of Roux on Nietzsche's thought include Gregory Moore (2002, 2006), Henry Staten (2006), and Robert Holub (2018).

Moore (2002) and James Pearson (2023) have argued that *Der Kampf der Theile* had a strong influence on Nietzsche's concept of will to power. In Nietzsche's vision of an internally divided self, different drives contend for power and influence, sometimes subduing or eliminating weaker drives, and by this means the self forges a higher unity. The similarities are strong but so are the differences. For Roux, the struggle occurred among material parts whereas for Nietzsche it was a struggle between intentional drives.

DATABASES VARY IN THE comprehensiveness of their coverage for different historical periods (and simple searches miss variant citations to the same work). Despite all these caveats, my bibliographic analysis agrees with Thomas Heams's (2012, 2016) conclusion that *Der Kampf der Theile* was a largely forgotten work through most of the twentieth century (except in the field of Nietzsche studies) but has been subject to renewed interest from the biological sciences in recent years. The reasons for this renewed interest are multiple, as are the reasons why you might be interested in reading our translation. Perhaps you are a historian of biology interested in evolutionary ideas in the nineteenth century or in the history of perennial arguments between neo-Lamarckians and neo-Darwinians. Perhaps you are an evolutionary biologist interested in the relation between evolution and development, or in the interaction of different levels of selection, or in the role of facilitated variation and phenotypic plasticity in adaptive evolution, or in functional self-organization or epigenetics. Perhaps you are a developmental biologist who knows of Roux as a founding father of experimental embryology. Perhaps you are a neuroscientist interested in his ideas on functional stimuli. Perhaps you are a physical anthropologist interested in what Roux has to say about the remodeling of bone by physical activity. Perhaps you are a systems biologist interested in the relations between parts and wholes of biological systems. Or perhaps you are a philosopher interested in Friedrich Nietzsche. Roux's book has something interesting to say to all these readers.

Whatever your area of interest, our intent in translating Roux's text is to make his book available for larger discourse in many fields.

Acknowledgments

The translation benefited from discussions with Judith Ryan, Peter Burghardt, and John Hamilton on the intricacies of German. The translation and historical research were greatly facilitated by Google's digitization of nineteenth-century books (including *Der Kampf der Theile*). The introductory and bibliographic essays have benefited from the advice of Janice Audet and Rachel Field at Harvard University Press.

Bibliography

Ackermann, C. T. 1884. *Mechanismus und Darwinismus in der Pathologie: Rede gehalten beim Antritt des Rectorates der Königlichen Friedrichs-Universität Halle-Wittenberg, am 12. Juli 1884.* Halle: Max Niemeyer.

———. 1894. "Die pathologische Bindegewebsneubildung [sic] in der Leber und Pflügers teleologisches Causalgcsctz." In *Festschriften der vier Fakultäten zum zweihundertjahrigen Jubiläum der Vereingten Friedrichs-Universität Halle-Wittenberg, den 3. August 1894: Festschrift der Medizinischen Fakultät,* 1–15. Halle: Waisenhauses.

Allen, G. E. 1975. *Life Science in the Twentieth Century.* New York: John Wiley.

Andler, C. 1928. *Nietzsche, sa vie et sa pensée. IV. La maturité de Nietzsche, jusqu'à sa mort.* Paris: Bossard.

———. 1958. *Nietzsche, sa vie et sa pensée.* 5th ed. Paris: Gallimard.

[Anonymous]. 1881. *Der Kampf der Theile im Organismus. Literarisches Centralblatt für Deutschland* 46: 1570–1571.

Baer, K., von. 1866. "Über Prof. Nic. Wagner's Entdeckung von Larven, die sich fortpflanzen, Herrn Ganin's verwandte und ergänzende Beobachtungen und über die Paedogenesis überhaupt." *Bulletin de l'Académie Impériale des Sciences de St-Pétersbourg* 9: 64–137.

Baer, K. E. von, and A. Dohrn. 1993. "Correspondence: Karl Ernst von Baer [1792–1876], Anton Dohrn [1840–1909]." Translated by C. Groeben and J. M. Oppenheimer. *Transactions of the American Philosophical Society* 83 (3).

Barfurth, D. 1891. Zur Regeneration der Gewebe. *Archiv für mikroskopische Anatomie* 37: 406–491.

Boeke, J. 1921. "The Innervation of Striped Muscle-Fibres and Langley's Receptive Substance." *Brain* 44: 1–22.

Boll, F. 1876. *Das Princip des Wachsthums. Eine anatomische Untersuchung.* Berlin: August Hirschwald.

Burnet, F. M. 1959. *The Clonal Selection Theory of Acquired Immunity.* Nashville, TN: Vanderbilt University Press.

Buss, L. 1987. *The Evolution of Individuality.* Princeton, NJ: Princeton University Press.

Butler, S. 1880. *Unconscious Memory: A Comparison between the Theory of Dr. Ewald Hering and the "Philosophy of the unconscious" of Dr. Edward von Hartmann.* London: David Bogue.

Canstatt, K. F. 1842. "Atrophie." In *Handwörterbuch der Physiologie: mit Rücksicht auf physiologische Pathologie,* vol. 1, edited by R. Wagner, 24–34. Braunschweig: Friedrich Vieweg.

Changeux, J. P., P. Courrège, and A. Danchin. 1973. "A Theory of Epigenesis of Neuronal Networks by Selective Stabilization of Synapses." *Proceedings of the National Academy of Sciences of the United States of America* 70 (10): 2974–2978.

Chernyak, L., and A. I. Tauber. 1990. "The Idea of Immunity: Metchnikoff's Metaphysics and Science." *Journal of the History of Biology* 23: 187–249.

Churchill, F. B. 1994. "Roux, Wilhelm." In *Dictionary of Scientific Biography.* 2nd ed., edited by C. C. Gillispie, 11:570–575. New York: Charles Scribner.

——. 2015. *August Weismann: Development, Heredity, and Evolution.* Cambridge, MA: Harvard University Press.

Claus, C. 1888. *Ueber die Wertschätzung der natürlichen Zuchtwahl als Erklärungsprincip.* Wien.

Cowin, S. C. 2001. "The False Premise of Wolff's Law." In *Bone Mechanics Handbook.* 2nd ed., edited by S. C. Cowin, chapter 30. Boca Raton, FL: CRC Press.

Cranefield, P. F. 1957. "The Organic Physics of 1847 and the Biophysics of Today." *Journal of the History of Medicine and Allied Sciences* 12: 407–423.

Darwin, C. 1859. *On the Origin of Species by Means of Natural Selection; Or the Preservation of Favoured Races in the Struggle for Life.* London: John Murray.

——. 1868. *The Variation of Animals and Plants under Domestication.* 2 vols. London: John Murray.

——. 1869. *On the Origin of Species by Means of Natural Selection; Or the Preservation of Favoured Races in the Struggle for Life.* 5th ed. London: John Murray.

———. 1876. *The Origin of Species by Means of Natural Selection; Or the Preservation of Favoured Races in the Struggle for Life.* 6th ed. London: John Murray.

———. 1881. *The Formation of Vegetable Mould through the Action of Worms, with Observations on Their Habits.* London: John Murray.

———. 1887. *The Life and Letters of Charles Darwin, Including an Autobiographical Chapter.* 3 vols., edited by F. Darwin. London: John Murray.

Delage, Y. 1895. *La structure du protoplasma et les théories sur l'hérédité et les grands problèmes de la biologie générale.* Paris: Reinwald.

Díaz, B., and E. Moreno. 2005. "The Competitive Nature of Cells." *Experimental Cell Research* 306: 317–322.

Dibbets, J. M. H. 1991. "One Century of Wolff's Law." In *Bone Dynamics in Orthodontic and Orthopedic Treatment*, edited by D. S. Carlson and S. A. Goldstein, 27:1–13. Ann Arbor: University of Michigan.

Du Bois-Reymond, E. 1881. *Über die Übung: Rede, zur Feier des Stiftungstages der Militär-Ärztlichen Bildungs-Anstalten, am 2. August 1881.* Berlin: August Hirschwald.

Edelman, G. M. 1987. *Neural Darwinism: The Theory of Neuronal Group Selection.* New York: Basic Books.

Exner, S. 1879. "Physiologie der Grosshirnrinde." In *Handbuch der Physiologie*, edited by L. Herman, vol 2, sect. 2, 189–350. Leipzig: Vogel.

Flemming ,W. 1880. "Über Epithelregeneration und sogenannte freie Kernbildung." *Archiv für mikroskopische Anatomie* 18: 347–364.

———. 1882. *Zellsubstanz, Kern und Zelltheilung.* Leipzig: Vogel.

Fraisse, P. 1885. *Die Regeneration von Geweben und Organen bei den Wirbelthieren, besonders Amphibien und Reptilien.* Cassel: T. Fischer.

Gawne, R., K. Z. McKenna, and M. Levin. 2020. "Competitive and Coordinative Interactions between Body Parts Produce Adaptive Developmental Outcomes." *BioEssays* 2020: 1900245.

Gerhart, J., and M. Kirschner. 2009. "The Theory of Facilitated Variation." *Proceedings of the National Academy of Sciences of the United States of America* 104: 8582–8589.

Gliboff, S. 2008. *H. G. Bronn, Ernst Haeckel, and the Origins of German Darwinism.* Cambridge, MA: MIT Press.

Gould, S. J. 2002. *The Structure of Evolutionary Theory.* Cambridge, MA: Harvard University Press.

Guldberg, G. A. 1895. *Om Darwinismen og dens Rækkevidde.* Kristiania: Dybwad.

Haeckel, E. 1866. *Generelle Morphologie der Organismen. Allgemeine Grundzüge der organischen Formen-Wissenschaft, mechanisch begründet durch die von Charles Darwin reformirte Descendenz-Theorie.* 2 vols. Berlin: G. Reimer.

————. 1868. *Natürliche Schöpfungsgeschichte: Gemeinverständliche wissenschaftliche Vorträge über die Entwicklungslehre im Allgemeinen und diejenige von Darwin, Goethe und Lamarck im Besonderen, über die Anwendung derselben auf den Ursprung des Menschen und andere damit zusammenhängende Grundfragen der Naturwissenschaft.* Berlin: G. Reimer.

————. 1876. *Die Perigenesis der Plastidule oder die Wellenzeugung der Lebenstheilchen. Ein Versuch zur mechanischen Erklärung der elementaren Entwickelungs-Vorgänge.* Berlin: Georg Reimer.

Hamburger, V. 1997. "Wilhelm Roux: Visionary with a Blind Spot." *Journal of the History of Biology* 30: 229–238.

Hartmann, E. von. 1883. "Eine neue Erweiterung des Darwinismus." *Der Gegenwart* 24 (40): 212–214.

Harvey, W. 1651. *Exercitationes de generatione animalium.*

Heams, T. 2012. "Selection within Organisms in the Nineteenth Century: Wilhelm Roux's Complex Legacy." *Progress in Biophysics and Molecular Biology* 110: 24–33.

————. 2016. "Préface: *La lutte des parties dans l'organisme,* ou l'impasse visionaire." In W. Roux, *La lutte des parties dans l'organisme,* edited by T. Heams and M. Silberstein, 26–30. Translated by L. Cohort, S. Danizet-Bichet, A. L. Pasoco-Saligny, and C. Thebault. Paris: Éditions Matériologiques.

Helmholtz, H. 1865. "Über das Verhältniss der Naturwissenschaften zur Gesammtheit der Wissenschaft." In *Populäre wissenschaftliche Vorträge von H. Helmholtz,* 1–29. Braunschweig: Friedrich Vieweg.

Hertwig, O. 1922. *Das Werden der Organismen. Zur Widerlung von Darwins Zufallstheorie durch das Gesetz in der Entwicklung.* 3rd ed. Jena: Gustav Fischer.

Holub, R. C. 2018. *Nietzsche in the Nineteenth Century.* Philadelphia: University of Pennsylvania Press.

Jacobson, M. 1970. *Developmental Neurobiology.* New York: Holt, Rinehart and Winston.

————. 1978. *Developmental Neurobiology.* 2nd ed. New York: Plenum.

————. 1991. *Developmental Neurobiology.* 3rd ed. New York: Plenum.

Jerne, N. K. 1955. "The Natural-Selection Theory of Antibody Formation." *Proceedings of the National Academy of Sciences of the United States of America* 41: 849–857.

Kant, I. (1790) 2000. *Critique of the Power of Judgment.* [*Kritik der Urteilskraft.*] Translated by P. Guyer and E. Matthews. Cambridge: Cambridge University Press.

Kellogg, V. L. 1907. *Darwinism To-day: A Discussion of Present-Day Scientific Criticism of the Darwinian Selection Theories, Together with a Brief Account of the*

Principal Other Proposed Auxiliary and Alternative Theories of Species-Forming. New York: Holt.

Kiermayer, A. 2020. "The Evolution of German Cut Fencing in the 19th Century Viewed through the Works of Friedrich August Wilhelm Ludwig Roux." *Acta Periodica Duellatorum* 6: 77–102.

Kupiec, J. J. 1986. "A Probabilist Theory for Cell Differentiation: The Extension of Darwinian Processes to Embryogenesis." *Speculations in Science and Technology* 9: 19–22.

———. 1997. "A Darwinian Theory for the Origin of Cellular Differentiation." *Molecular and General Genetics* 255: 201–208.

Kurz, H., K. Sandau, and B. Christ. 1997. "On the Bifurcation of Blood Vessels— Wilhelm Roux's Doctoral Thesis (Jena 1878)—A Seminal Work for Biophysical Modelling in Developmental Biology." *Annals of Anatomy* 179 (1): 33–36.

Lee, T. C., and D. Taylor. 1999. "Bone Remodelling: Should We Cry Wolff?" *Irish Journal of Medical Science* 168: 102–105.

Lenoir, T. 1981. "Teleology without Regrets: The Transformation of Physiology in Germany: 1790–1847." *Studies in the History and Philosophy of Science* 12: 293–354.

Levin, M. 2021. "Life, Death, and Self: Fundamental Questions of Primitive Cognition Viewed through the Lens of Body Plasticity and Synthetic Organisms." *Biochemical and Biophysical Research Communications* 564: 14–133.

Maienschein, J. 1994. "The Origins of *Entwicklungsmechanik*." In *A Conceptual History of Modern Embryology,* edited by S. F. Gilbert, 43–61. New York: Plenum.

Mayer, S. 1879. *Specielle Nervenphysiologie. Handbuch der Physiologie des Nervensystems,* special volume, *Handbuch der Physiologie,* edited by L. Hermann, 209–212. Leipzig: Vogel.

Mayr, E. 1982. *The Growth of Biological Thought: Diversity, Evolution, and Inheritance.* Cambridge, MA: Belknap Press.

———. 1983. How to carry out the adaptationist programme? *American Naturalist* 121: 324–334.

———. 1992. Controversies in retrospect. *Oxford Surveys in Evolutionary Biology* 8: 1–34.

Merkel, F. 1882. [Review of *Der Kampf der Theile im Organismus*]. *Jahresberichte über die Leistungen und Fortschritte in der Gesammten Medicin* 16 (vol. 1): 118.

Metchnikoff, E. 1892. "La lutte pour l'existence entre des diverses parties de l'organisme." *Revue scientifique* 50: 321–326.

Michaelson, J. 1987. "Cell Selection in Development." *Biological Reviews of the Cambridge Philosophical Society* 62 (May): 115–139.

Mocek, R. 1974. *Wilhelm Roux, Hans Driesch: zur Geschichte der Entwicklungsphysiologie der Tiere ("Entwicklungsmechanik")*. Jena: Gustav Fischer.

Moore, G. 2002. *Nietzsche, Biology and Metaphor*. Cambridge: Cambridge University Press.

———. 2006. *Nietzsche and Evolutionary Theory*, edited by K. Ansell Pearson, 517–531. Oxford: Blackwell.

Morange, M. 2012. "What History Tells Us, XXVIII. What Is Really New in the Current Evolutionary Theory of Cancer?" *Journal of Biosciences* 37: 609–612.

Müller-Lauter, W. 1978. "Der Organismus als innerer Kampf. Der Einfluss von Wilhelm Roux auf Friedrich Nietzsche." *Nietzsche-Studien* 7: 189–235.

———. 1999. "The Organism as Inner Struggle: Wilhelm Roux's Influence on Nietzsche." In *Nietzsche: His Philosophy of Contradictions and the Contradictions of His Philosophy*, 161–182. Translated by D. J. Parent. Urbana: University of Illinois Press.

Nägeli, C. W. von. 1884. *Mechanisch-physiologische Theorie der Abstammungslehre*. Munich: R. Oldenbourg.

Nietzsche, F. 1906. *Der Wille zur Macht, 1884/1888. Versuch einer Umwerthung aller Werthe*. Leipzig: Naumann.

———. 2017. *The Will to Power: Selections from the Notebooks of the 1880s*, edited by R. K. Hill. Translated by R. K. Hill and M. A. Scarpitti. London: Penguin Classics.

Nothnagel, H. 1894. *Die Anpassung des Organismus bei pathologischen Veränderungen: Vortrag, gehalten beim XI. internationalen medizinischen Congress in Rom am 31. März 1894*. Wien: L. Bergmann.

Nyhart, L. K. 1995. *Biology Takes Form: Animal Morphology and the German Universities 1800–1900*. Chicago: University of Chicago Press.

Oppenheimer, J. M. 1967. *Essays in the History of Embryology and Biology*. Cambridge, MA: MIT Press.

Pauly, A. 1905. *Darwinismus und Lamarckismus. Entwurf einer psychophysischen Teleologie*. Munich: Ernst Reinhardt.

Pearson, J. S. 2023. "Nietzsche on the Necessity of Repression." *Inquiry* 66: 1–30.

Pfeffer, G. 1893. "Die Umwandlung der Arten, ein Vorgang funktioneller Selbstgestaltung." *Verhandlungen des Naturwissenschaftlichen Vereins in Hamburg* 3 (1): 44–87.

Pflüger, E. F. W. 1877. "Die teleologische Mechanik der lebendigen Natur." *Archiv für die gesamte Physiologie des Menschen und der Tiere* 15: 57–103.

Pichot, A. 2003. "Mémoire pour rectifier les jugements du public sur la révolution biologique." *Esprit* 297(8/9): 104–110.

Plate, L. 1899. "Die Bedeutung und Tragweite des Darwin'schen Selectionsprincips." *Verhandlungen der deutschen zoologischen Gesellschaft* 9: 59–208.

———. 1913. *Selectionsprinzip und Probleme der Artbildung. Ein Handbuch des Darwinismus.* 4th ed. Leipzig: Wilhelm Engelmann.

Preyer, W. T. 1879. "Die Concurrenz in der Natur." *Nord und Süd* 8 (January–March): 191–212.

———. 1880. "Die Concurrenz in der Natur." In *Naturwissenschaftliche Thatsachen und Probleme: Populäre Vorträge,* 65–96. Berlin: Gebrüder Pætel.

———. 1882. *Die Concurrenz in der Natur.* Breslau: S. Schottlaender.

Reichenau, W. von. 1881–1882. "Über den Ursprung der männlichen sekundären Geschlechtscharaktere, insbesondere bei den Blatthornkäfern." *Kosmos* 10: 172.

Reynolds, A. 2007. "The Theory of the Cell State and the Question of Cell Autonomy in Nineteenth Century and Early Twentieth-Century Biology." *Science in Context* 20: 71–95.

Riehl, A. 1876. *Der philosophische Kritizismus und seine Bedeutung für die positive Wissenschaft.* 3 vols. Leipzig: Wilhelm Engelmann.

Roesler, H. 1981. "Some Historical Remarks on the Theory of Cancellous Bone Structure (Wolff's Law)." In *Mechanical Properties of Bone,* edited by S. C. Cowin, 27–42. New York: American Society of Mechanical Engineers.

Romanes, G. J. 1881a. "The Struggle of Parts in the Organism." *Nature* 24: 505–506.

———. 1881b. "The Formation of Vegetable Mould." *Nature* 24: 553–556.

———. 1881c. "Unconscious memory, &c." *Nature* 23: 285–287.

———. 1888. "Lamarckism versus Darwinism." *Nature* 38: 413.

———. 1895. *Darwin, and after Darwin. II. Post-Darwinian Questions: Heredity and Utility,* edited by C. L. Morgan. Chicago: Open Court.

———. 1896. *The Life and Letters of George John Romanes, M.A., LL.D., F.R.S.* London: Longmans, Green.

Roux, W. 1878. "Uber die Verzweigungen der Blutgefässe." PhD diss., Jena.

———. 1879. "Über die Bedeutung der Ablenkung des Arterienstammes bei der Astabgabe." *Jenaische Zeitschrift für Naturwissenschaft* 13: 321–337.

———. 1881a. *Der Kampf der Theile im Organismus: Ein Beitrag zur Vervollständigung der mechanischen Zweckmässigkeitslehre.* Leipzig: Engelmann.

———. 1881b. "Der Kampf der Theile im Organismus." *Biologisches Centralblatt* 1: 241–251.

———. 1885. "Beiträge zur Morphologie der functionellen Anpassung. 3. Beschreibung und Erläuterung einer knöchernen Kniegelenksankylose (des Trockenpräparates

Nr. 691 der pathologisch-anatomischen Sammlung zu Würzburg)." *Archiv für Anatomie und Entwickelungsgeschichte* 120–158.

———. 1893. "Das Gesetz der Transformation der Knochen. I. Theoretischer Theil." *Berliner klinische Wochenschrift* 30: 509–511, 533–535, 557–558.

———. 1895a. "Der züchtende Kampf der Theile im Organismus oder die 'Theilauslese' im Organismus. Zugleich eine Theorie der 'functionellen Anpassung.'" In *Gesammelte Abhandlungen über Entwickelungsmechanik der Organismen*, Abhandlungen I–XII, *Vorwiegend über functionelle Anpassung*. vol. 1, 135–422. Leipzig: Wilhelm Engelmann.

———. 1895b. "Der züchtende Kampf der Theile im Organismus. Autoreferat über Nr. 4." In *Gesammelte Abhandlungen über Entwickelungsmechanik der Organismen*, Abhandlungen I–XII, *Vorwiegend über functionelle Anpassung*, vol. 1, 423–437. Leipzig: Wilhelm Engelmann.

———. 1923a. "Wilhelm Roux in Halle a. S." In *Die Medizin der Gegenwart in Selbstdarstellungen*. 2 vols., edited by L. R. Grote, 1: 141–206. Leipzig: Felix Meiner.

———. 1923b. "Prinzipielles der Entwickelungsmechanik." *Annalen der Philosophie* 3: 454–473.

———. 2016. *La lutte des parties dans l'organisme*, edited by T. Heams and M. Silberstein. Translated by L. Cohort, S. Danizet-Bichet, A. L. Pasoco-Saligny, and C. Thebault. Paris: Éditions Matériologiques.

Russell, E. S. 1916. *Form and Function: A Contribution to the History of Animal Morphology*. London: John Murray.

Schiller, F. 1880. *Wallenstein: ein dramatisches Gedicht von Schiller*. Stuttgart: Cotta.

Spencer, H. 1855. *The Principles of Psychology*. London: Longman, Brown, Green, and Longmans.

———. 1860. "The Social Organism." *Westminster Review* 73: 90–121.

———. 1865. *The Principles of Biology*. 2 vols. London: Williams and Norgate.

———. 1876. *The Principles of Sociology*. 3 vols. London: Williams and Norgate.

———. 1894. "Weismannism Once More." *Contemporary Review* 66: 592–608.

Spitta, H. 1881. [Review of *Der Kampf der Theile im Organismus*]. *Deutsche Literaturzeitung* 2: 1148.

Spitzer, H. 1886. *Beiträge zur Descendenztheorie und zur Methodologie der Naturwissenschaft*. Leipzig: F. A. Brockhaus.

Staten, H. 2006. "A Critique of the Will to Power." In *A Companion to Nietzsche*, edited by K. Ansell Pearson, 565–582. Oxford: Blackwell.

Strasser, H. 1883. *Zur Kenntniss der funktionellen Anpassung der quergestreiften Muskeln. Beiträge zu einer Lehre von dem kausalen Zusammenhang in den Entwicklungsvorgängen des Organismus.* Stuttgart: F. Enke.

Swiatczak, B. 2023. "Evolution within the Body: The Rise and Fall of Somatic Darwinism in the Late Nineteenth Century." *History and Philosophy of the Life Sciences* 45: 8.

Tornier, G. 1884. *Der Kampf mit der Nahrung: Ein Beitrag zum Darwinismus.* Berlin: Wilhelm Issleib.

Unbehaun, J. 1896. *Versuch einer philosophischen Selektionstheorie.* Jena: Gustav Fischer.

Vasquez Kuntz, K. L., S. A. Kitchen, T. L. Conn, S. A. Vohsen, A. N. Chan, M. J. A. Vermeij, C. Page, K. L. Marhaver, and I. B. Baums. 2022. "Inheritance of Somatic Mutations by Animal Offspring." *Science Advances* 8: eabn0707.

Virchow, R. 1859. *Die Cellularpathologie in ihrer Begründung auf physiologische und pathologische Gewebelehre.* 2nd ed. Berlin: August Hirschwald.

———. 1863. "Über Erblichkeit. I. Die Theorie Darwin's." *Deutsche Jahrbücher für Politik und Literatur* 6 (January–March): 339–358.

———. 1880. "Krankheitswesen und Krankheitsursachen." *Virchows Archiv für pathologische Anatomie und Physiologie und für klinische Medizin* 79: 1–19, 185–228

Vogel, J. 1844. "Hypertrophie." In *Handwörterbuch der Physiologie: mit Rücksicht auf physiologische Pathologie,* vol. 2., edited by R. Wagner, 186–191. Braunschweig: Friedrich Vieweg.

Volkmann, A. W. 1874. Zur Entwickelung der Organismen. *Bericht über die Sitzungen der naturforschenden Gesellschaft zu Halle* (5 July) 27–36.

Weikart, R. 1993. "The Origins of Social Darwinism in Germany, 1859–1895." *Journal of the History of Ideas* 54: 469–488.

Weismann, A. 1891. "On Heredity, 1883." In *Essays upon Heredity and Kindred Biological Problems,* vol. 1, translated by A. E. Shipley, edited by E. B. Poulton, S. Schönland, and A. E. Shipley, 67–106. Oxford: Clarendon.

———. 1892. *Das Keimplasma: Eine Theorie der Vererbung.* Jena: Gustav Fischer.

———. 1893. *The Germ-Plasm: A Theory of Heredity.* Translated by W. N. Parker and H. Rönnfeldt. New York: Charles Scribner.

———. 1894. *The Effect of External Influences upon Development.* London: Henry Frowde.

———. 1895. *Neue Gedanken zur Vererbungsfrage: Eine Antwort an Herbert Spencer.* Jena: Gustav Fischer.

———. 1896. "Germinal Selection." *Monist* 6: 250–293.

——. 1902. *Vorträge über Deszendenztheorie.* Jena: Fischer.

——. 1904. *The Evolution Theory.* Translated by J. A. Thomson and M. R. Thomson. London: Edward Arnold.

West-Eberhard, M. J. 2003. *Developmental Plasticity and Evolution.* Oxford: Oxford University Press.

Wolff, J. 1892. *Das Gesetz der Transformation der Knochen.* Berlin: Hirschwald.

——. 1899. "Die Lehre von der functionellen Knochengestalt." *Archiv für pathologische Anatomie und Physiologie* 155: 256–315.

Wundt, W. 1881. *Logik. Eine Untersuchung der Principien der Erkenntnis und der Methoden wissenschaftlicher Forschung,* vol. 2. *Methodenlehre.* Stuttgart: Ferdinand Enke.

Yanai, I., and M. Lercher. 2020. "Renaissance Minds in 21st Century Science." *Genome Biology* 21: 67.

Index

soul (*Seel*), 176, 208, 218, 222n7. *See also* spirit; will

Speck, Karl, 100

Spencer, Herbert, 29, 88–89, 91, 93; definition of life of, 207; dispute with Weismann, August, 240; functional hyperemia and, 29, 147n5; indirect equilibration and, 2, 29

spinal cord, 48, 52, 135–137, 139, 143; atrophy of nerves, 163; injury to, 71, 107, 117–118, 137, 140–141; trophic stimulus for muscles, 71–72, 137, 143, 182. *See also* ganglion cells

spirit (*Geist*), 52, 54, 60n12, 164, 167. *See also* soul; will

Spitta, Heinrich, 31

Spitzer, Hugo, 31

spleen, 50, 120, 152, 160, 175

stimulus: abnormal or inappropriate, 136, 178, 201, 249; chemical, 142, 145, 152, 160, 173, 175; definitions of, 99n13, 210; electrical, 133, 139, 143, 164; frequency and intensity of, 99, 103, 129–132, 177, 179, 180; mechanical, 101, 145, 152, 164, 175, 185; psychic, 191, 227; sensory, 131, 142, 173–174; in struggle of cells, 108–109

stimulus, functional (*functionell Reiz*): of connective tissue, 142, 188n3; definitions, 99, 129, 210; direct or indirect, 125, 143, 164; as epigenesis, 205–206; and functional harmony, 198; indispensable life stimulus, 102, 125, 130, 176–179, 218, 220; qualitative (differentiating) effect, 103, 131, 168–171; quantitative (formative) effect, 114, 130–131, 180–184, 191–192, 201, 220; and self-organization, 194–195, 231; trophic effect of, 22, 100, 170–171, 176; trophic nerves and, 138; of tumors, 73, 144–146, 249

stimulus life versus embryonic life, 70–71, 73–74, 179–180, 196–198, 232; as contrast between epigenesis and evolution, 74n15, 203n8. *See also* embryogenesis

Stirling, William, 151

Strasser, Hans, 32, 55, 205

stress, 189, 206

struggle (*Kampf*), meaning, 12–13, 217

struggle for existence. *See* struggle of individuals

struggle for food, 28n1, 91, 96, 101, 121, 123–124; Rudolf Virchow, 112

struggle for space, 82, 126, 179, 247; balance among organs, 119–121; territorial expansion, 115; by unequal growth, 93–98; by unequal pressure, 111–113, 124

struggle for stimuli, 101

struggle of cells, 107–109, 112–114, 230

struggle of individuals, 104–106, 123–126, 174, 202–204, 219–220, 230–232; struggle among persons, 95. *See also* levels of selection; natural selection; selection, personal

struggle of living molecules, 15, 92–100, 103–106, 230

struggle of organs, 88–89, 119–122. *See also* harmony of parts

struggle of parts (*Kampf der Theile*), 12–13, 81, 165; qualitative inequality and, 86; struggle of dissimilar parts, 115–116, 119–120, 126; struggle of similar parts, 29, 123–124; struggles of similar and dissimilar parts, comparisons, 88–89, 91, 121, 230, 238. *See also* harmony of parts; selection of parts

Struggle of Parts (Roux). See *Kampf der Theile* (Roux)

struggle of tissues, 115–117

Swiatczak, Bartlomiej, 243–244

syphilis, 79, 107, 115, 145

Tauber, Alfred, 247

teeth, 120, 181

teleology, 9–10, 40, 166; dualism, 60; dysteleology, 42; Eduard Pflüger, 33n4, 223; Immanuel Kant, 8–9; lastingness, 38; mechanical, 199n6; psychophysical, 242; purposeful will, 202, 232; teleophobia, 9; Wilhelm Wundt, 31, 82. *See also* purposiveness (*Zweckmäßigkeit*)

tendons, 71–72; effects of will, 202; response to tension, 48, 50, 56, 114. *See also* tension

tension, 55–56, 128, 130, 199, 209; blood vessels and, 163, 193; bone and, 55, 130–132, 139, 231; connective tissue and, 49, 129. *See also* tendons